Sustainable development of the Ganges-Brahmaputra-Meghna basins

Edited by Asit K. Biswas and Juha I. Uitto

TOKYO • NEW YORK • PARIS

United Nations University Press
The United Nations University, 53-70, Jingumae 5-chome,
Shibuya-ku, Tokyo, 150-8925, Japan
Tel: +81-3-3499-2811 Fax: +81-3-3406-7345
E-mail: sales@hq.unu.edu
http://www.unu.edu

United Nations University Office in North America
2 United Nations Plaza, Room DC2-1462-70, New York, NY 10017, USA
Tel: +1-212-963-6387 Fax: +1-212-371-9454
E-mail: unuona@igc.apc.org

United Nations University Press is the publishing division of the United Nations University.

Cover design by Joyce C. Weston
Cover photograph by Manisha Verma

Printed in the United States of America

UNUP-1041
ISBN 92-808-1041-3

Library of Congress Cataloging-in-Publication Data

Sustainable development of the Ganges-Brahmaputra-Meghna basins /
edited by Asit K. Biswas and Juha I. Uitto.
p. cm.
Includes bibliographical references and index.
ISBN 92-808-1041-3
1. Water resources development-Ganges River Watershed (India and Bangladesh) 2. Sustainable development-Ganges River Watershed (India and Bangladesh) I. Biswas, Asit K. II. Uitto, Juha I. III. Title.
HD1698.G36 S87 2000
333.91′15′09541—dc21
00-011973

Sustainable development of the Ganges

The UNU Programme on Integrated Basin Management focuses on water management, approaching the complex problematique from three particular angles: governance, capacity-building, and management tools. The programme is carried out through field-based research encompassing both natural and social sciences. It utilizes extensive networks of scholars and institutions in both developing and industrialized countries. This work is intended to contribute to policy-making by the United Nations and the international community, as well as to capacity-building in developing countries.

The Water Resources Management and Policy series disseminates the results of research carried out under the Programme on Integrated Basin Management and related activities. The series focuses on policy-relevant topics of wide interest to scholars, practitioners, and policy-makers.

Earlier books in this series are:

Hydropolitics Along the Jordan River: Scarce Water and its Impact on the Arab-Israeli Conflict by Aaron T. Wolf

Managing Water for Peace in the Middle East: Alternative Strategies by Masahiro Murakami

Freshwater Resources in Arid Lands edited by Juha I. Uitto and Jutta Schneider

Central Eurasian Water Crisis: Caspian, Aral, and Dead Seas edited by Iwao Kobori and Michael H. Glantz

Latin American River Basins: Amazon, Plata, and São Francisco edited by Asit K. Biswas, Newton V. Cordiero, Benedito P. F. Braga, and Cecilia Tortajada

Water for Urban Areas: Challenges and Perspectives edited by Juha I. Uitto and Asit K. Biswas

Transboundary Freshwater Dispute Resolution: Theory, Practice, and Annotated References by Heather L. Beach, Jesse Hamner, J. Joseph Hewitt, Edy Kaufman, Anja Kurki, Joe A. Oppenheimer, and Aaron T. Wolf

Contents

List of tables and figures

Note on measurements

Both metric and imperial measurements have been used in this book, in accordance with the systems in common usage in the countries concerned.

Preface

Asit K. Biswas and Juha I. Uitto

Water has always been regarded as a key resource and central for economic development in South Asia. A growing population and accelerating economic development activities in the region have made sustainable water management increasingly critical. The Ganges-Brahmaputra-Meghna basin now contains some 550 million people, more than the countries of North America and Western Europe combined. It also contains the largest concentration of the world's poor, more than all of sub-Saharan Africa. The region, however, has considerable development potential, especially if its natural resources could be harnessed sustainably, efficiently, and rationally.

The sharing of the water resources of the Ganges-Brahmaputra-Meghna river system has long been a subject of dispute among the four countries which share the basin: Bangladesh, Bhutan, India, and Nepal. This book presents a unique view of the potential development options for the shared waters as perceived by leading policy-makers and experts from the countries themselves.

The issues related to the sharing of the South Asian waters must be understood against the natural, historical, political, social, and economic backdrop. Low-lying Bangladesh is the last country in line for Ganges water. The river has its beginning in the Himalayas in Nepal, then flows through India before reaching Bangladesh and draining into the Bay of Bengal. The basin of the Brahmaputra is largely in Assam, the Indian state in the far north-east, all but cut off from the rest of the country, and receives much of its waters from the small upstream kingdom of Bhutan.

The Ganges, one of the world's mighty rivers, no longer reaches the sea every year. The upstream diversions and other water demands do not leave enough water for the river to reach its natural outlet in the Bay of Bengal. Since the construction of the Farakka barrage to divert water to Calcutta from the Indian Ganges in the mid-1970s, sharing of water downstream has become a highly contentious political issue between Bangladesh and India. The case highlights the crucial role water plays in the economic development of the region and how it affects all sectors. The rapid growth of the Calcutta mega-city has brought with it heavy demand for water resources. Similarly, the Calcutta harbour requires sufficient water to allow large ships to moor there. The resulting seasonal water deficit poses development constraints on downstream Bangladesh. The lack of fresh water flowing out to the sea has caused a rapid advance of a saline front across the western portion of the river delta, threatening agricultural production.

For more than three decades, the development of the Ganges-Brahmaputra river system has been a hydropolitical bone of contention in the region. Issues arising after the 1971 war that resulted in the creation of Bangladesh still remain unresolved. An agreement between India and Bangladesh to share the dry-season flow from the Ganges expired in 1988. Since then, there has been little progress in the two countries agreeing to a plan to share the water. However, two recent treaties on the Ganges (between Bangladesh and India) and the Mahakali (between India and Nepal) have dramatically changed the political atmosphere in terms of water management.

If the total annual flow is considered, the Ganges has abundant water resources. The main problem is water scarcity during the lean season – from January to April – which affects both India and Bangladesh. Only 5 per cent of the total annual flow of the river would be sufficient to solve the downstream scarcity problems during the lean season. This would, however, require finding new ways to store water in catchments to be used when the flow diminishes. The possibility of flow augmentation should thus receive high priority in management of the water resources of the region.

While the lean-season scarcity problem is easy to resolve conceptually, problems arise when attempting to select a way that would be acceptable to all countries concerned. Historically, Bangladesh has preferred to have reservoirs constructed upstream, in Nepal. India, on the other hand, has supported the construction of a 300-km-long canal – through Bangladesh territory – to transfer water from the relatively water-rich Brahmaputra to the water-scarce Ganges basin. The plan has not met with favour in Bangladesh for a variety of reasons. It would involve resettling a large number of people in the country, and many serious environmental and social problems would probably arise. A canal through Bangladesh which could also be used for navigation from one part of India to another could have serious socio-political implications for both countries.

Against this complex historical and political background, the Untied Nations University (UNU) and the International Water Resources Association (IWRA), supported by the government of the Netherlands, convened a Ganges forum in March 1998. The main purpose of the Ganges-Brahmaputra forum was to provide the countries concerned with an independent platform where senior policy-makers and experts could quietly and objectively explore genuine potential for cooperation in the sustainable development of the water, land, and biotic resources, and come to understand and appreciate each others' resource needs. Groundwork for organizing the forum had been carefully prepared, starting with the joint UNU-IWRA Asian Water Forum organized in January 1995 (Biswas and Hashimoto 1996).

Participation in this very high-level but low-key forum was strictly restricted, by invitation only, to less than 30 senior policy-makers and experts from the basin countries. They were all invited in their personal capacities to allow for free and frank exchange of ideas, opinions, and facts on this complex hydropolitical issue. Because of the position of the participants, the meeting refrained from drawing any formal conclusions or making any specific recommendations, but rather outlined the options available. The role of the UNU and IWRA was to act as impartial facilitators and honest brokers in the process. This book reports on the views of the participants, providing rare insight into the thinking amongst the leading persons concerned with water resources development in these four countries.

Eight background papers were specifically commissioned for the forum. These analysed the historical background of the water-use conflicts, as well as the collaborations among the four basin countries; the future potential for water resources development in the basin; considerations of flow augmentation in the lean season; review of the social, environmental, and political conditions; and discussion of possible legal and institutional frameworks for future collaboration between the countries concerned.

A major problem identified in the forum was the apparent lack of information-sharing. Water experts in one country had surprisingly little access to the information from other basin countries. Even information freely available in one country is often not available in others. The present book also aims to contribute to information sharing. Although preliminary reports on the forum have been published earlier (Biswas, Nakayama, and Uitto 1998), this is the first time the full papers presented at the forum are made available.

The forum also resulted in an agreement to develop a sustainable development framework for the Ganges-Brahmaputra-Meghna region. Two eminent experts each were selected from Bangladesh, India, and Nepal following extensive consultations with the respective governments, water experts, NGOs, and external support agencies. The work of this group is now under way. It is hoped that this work will lead to concrete results for sustainable management of water and natural resources in this important region.

We are most grateful to the Ministry of Foreign Affairs, Government of the Netherlands, and especially to Mr A. J. Diphoorn, for the financial support received which made this book possible.

REFERENCES

Biswas, A. K. and T. Hashimoto, eds. 1996. *Asian International Waters: From Ganges-Brahmaputra to Mekong*. Delhi: Oxford University Press.

Biswas, A. K., M. Nakayama and J. I. Uitto. 1998. "Standing in Line for Water: Cooperation on the Ganges and Brahmaputra", *Work in Progress*, Vol. 15, No. 2, pp. 4–5. Tokyo: United Nations University Press.

1

Management of international waters: Opportunities and constraints

Asit K. Biswas

Introduction

Historically, global water demands have increased steadily with population growth and the subsequent increases in various types of human activities. With a steadily increasing world population, and mankind's eternal quest for higher and higher standards of living, there is no doubt that the demands on our natural resources, both non-renewable and renewable, will continue to increase well into the foreseeable future. Water, a renewable resource, will be no exception to this general trend.

Even when the global population stabilizes, the demands for resources like water are likely to continue to increase because of higher per capita demands from people in the developing world attaining better standards of living, as well as due to the changing lifestyles in both developed and developing countries. These trends are now clearly visible in countries like India, where already some 100 million people have reached a middle-class standard of living. This rapidly emerging class is slowly flexing its new-found political and economic muscles, and, in the area of water, unlike earlier generations, it is unlikely to remain satisfied with the *status quo* of a few hours' intermittent supply of dubious quality every day. The availability of adequate quantity and appropriate quality of water to an increasingly urban population in the developing world is likely to be an important political and social issue in the coming decades in most countries.

Three other factors should also be noted: water requirements for agricultural production and energy generation, and increasing water contamination due to accelerated human activities. Efficient irrigated agriculture is essential for ensuring reliable food production in the twenty-first century. At present nearly 55 per cent of all rice and wheat produced in the world comes from irrigated areas and some 2.4 billion people currently depend on irrigated agriculture for food, income, and employment. Current estimates indicate that 80 per cent of the additional food supplies required to feed the future world population will depend on irrigation (IIMI 1992). Reliable availability of an adequate quantity and quality of water for increasing agricultural production will continue to be an important factor well into the twenty-first century.

While water requirements for increasing total global agricultural production have received considerable attention in recent years (for example at the World Food Summit in Rome in 1996), water needs for energy production have been basically neglected thus far. High development and economic growth rates can be achieved, especially in the developing world, only if adequate energy is available. No large-scale electricity generation is possible without large quantities of water. In addition to hydropower generation, construction and operation of new thermal and nuclear plants would require significant additional quantities of cooling water, a fact that has basically escaped both water and energy planners thus far.

A good example of the implications of massive increases in electricity generation for national and international water resources is the Asian developing countries. These countries had a total generating capacity of 250,000 MW in 1990, nearly 70 per cent of which was thermally generated (mainly coal), with the balance of 30 per cent being accounted for mainly by hydropower. It was estimated that another 240,000 MW would be needed by the year 2000 to fulfil their current development plans. This near doubling of electricity requirements within only one decade means that the water needs of a rapidly expanding energy sector can no longer be ignored, especially as similar growth rates are expected to continue during the early part of the twenty-first century (Biswas and Hashimoto 1996). Similarly, in Turkey electricity demands are now growing by 7–8 per cent annually. It should also be noted that for England and Wales, some 36 per cent of all water abstracted at present is accounted for by the energy generation industry alone. The corresponding figure for France is much higher, and for Mexico it is 69 per cent. The future global water requirements for the energy sector thus need to be carefully considered.

Another important issue is increasing water contamination, especially in the developing world, which means that many sources of water must now receive expensive treatment before they can be used beneficially.

While much rhetoric can be noted in terms of water pollution control and ecosystems' conservation, in reality appropriate remedial actions on the ground are basically missing. Inadequate treatment of domestic and industrial wastewater (for example, in Mexico, only about 6 per cent of the total wastewater produced is properly treated) and complete neglect of non-point sources mean that water quality management urgently needs more attention than the lip service it is receiving at present.

These three issues, as well as other associated factors, mean that sustainable water management in the coming decades will face a challenge the like of which has not been seen before. In spite of the gravity of the situation, unfortunately we currently do no even have reasonably reliable assessments of the global situation in terms of water quantity and quality and associated factors. This fact is a damming indictment of international programmes like UNESCO's International Hydrological Programme, which after spending hundreds of millions of dollars over the decades has been unable to produce even such basic information. Similarly, the United Nations system and the Stockholm Environment Institute (United Nations 1997) recently published a report entitled *Comprehensive Assessments of the Freshwater Resources of the World*. While this report makes some useful points, it is most remarkable for the absence of any serious assessments of the global water quantity and quality conditions. In addition, no major country has prepared realistic estimates of future water demand and availability situations based on expected patterns of development, social and environmental requirements, economic considerations, changes in management techniques, involvement of the private sector, and other related issues. All these would be important considerations in the twenty-first century.

These factors and other related issues, when considered together, indicate the following.

- Water requirements in arid and semi-arid areas will continue to increase steadily in the foreseeable future for a variety of reasons.
- All easily available exclusively national sources of water have already been developed or are in the process of development, which means that the real costs of new projects in terms of per cubic metre of new water available will continue to increase. An analysis of domestic water supply projects supported by the World Bank (1992) indicates that the cost per cubic metre of water for the next generation of projects is often two or three times higher than for the present generation.
- Water needs for ecosystem and habitat preservation will increasingly be considered as a "legitimate" use in the future, probably within a decade, thus putting additional pressure on water available for existing "traditional" uses.
- For environmental and social reasons, the next generation of water

projects will take a significantly longer time to develop than currently anticipated by the planners, which will further intensify the anticipated water shortages in the foreseeable future.
- Water planning and management practices are likely to improve only incrementally in the near to medium term: the radical changes in such practices needed to resolve the water crisis appear to be highly unlikely to take place.

Importance of international waters

All the above factors will probably contribute to tremendous socio-political pressure to develop new sources of water. Since exclusively national and new sources of water which could be efficiently developed techno-economically are generally no longer available, there will be tremendous pressure in many countries to develop international water bodies – that is those rivers, lakes, and aquifers that are shared by two or more countries. This is because international bodies are often the only sources of water that can be developed economically. These water bodies were not considered for development in the past because of the absence of any agreements between the relevant countries on their utilization. The political risks and complexities were considered to be too high for their unilateral development by only one country. However, as water shortages in individual nations become more and more severe, and if and when they create serious internal political and social tensions and unrest, some countries may decide that it is worth the external political risk to develop such resources, even though it could imply a "beggar thy neighbour" attitude.

Over the past two decades, there has been an increasing number of situations where countries have built dams and barrages on the main stems of international rivers (sometimes even very close to their borders), and/or on major tributaries, which will affect the flow regime in downstream countries. Furthermore, the number of studies that have already been completed, or are under preparation, on the development of the major tributaries of international rivers, especially when they are under exclusive national jurisdictions, is increasing all the time. All such developments on major tributaries would clearly have perceptible impacts on the main international river. A good recent example of this was on the Mekong river. On 5 April 1995, plenipotentiaries from Cambodia, the Lao People's Democratic Republic, Thailand, and Viet Nam signed an agreement on cooperation for the sustainable development of the lower Mekong river basin at Chiang Rai, Thailand. Following the signing ceremony, the VIPs embarked on a boat tour of the Mekong river. The boat, however, was grounded in the river because, unknown to the participants, China was filling up a major new reservoir upstream. China is not a member of the newly constituted Mekong River Commission. Currently several dams are under construction and/or active consideration upstream in the Chinese territory, and China is the most upstream country in the Mekong river system.

International funding agencies have in the past generally declined to provide loans for the development of international waters until and unless the countries concerned have signed a mutually acceptable agreement. Without external financial assistance, developing countries have often been unable to construct capital-intensive water development projects on international rivers unilaterally. An analysis of the latest trends indicates that this situation appears to be changing, for the following reasons.

- Many of the countries concerned are now capable of raising the necessary investment funds from internal national sources.
- Private-sector funds, both international and national, can now be harnessed for such developments.
- International funding agencies appear to be taking a somewhat more liberal approach recently to supporting such development activities. For example, while they are still continuing to decline to finance construction of structures like the dams and hydropower plants, they have began to support associated activities like agricultural development, even though it would use water from a dam or hydropower project.
- In many cases international financing supports only a limited part of the total project cost. Absence of international funds could thus mean a simple extension of the time needed for construction: it may not necessarily stop the project.

Herein lies one of the principal challenges to the water profession in the twenty-first century: how to develop and manage the various international water sources sustainably and efficiently in full agreement and cooperation between the concerned countries so that the result can be a "win-win" situation for all parties. Development considerations will no longer be confined to techno-economical factors alone, as is generally the case for exclusively national projects; other factors will have to be considered, including bi-national or multinational political considerations, the power of the country in which the water development would take place compared to other co-basin countries, the importance of maintaining good relations between the countries concerned, and the general international and media interest in the project. These and other associated issues are likely to complicate water management processes in the future by several orders of magnitude. Hydropolitics is thus likely to become an increasingly important global issue in the future management of international river and lake basins and aquifers (Biswas *et al.* 1999).

Magnitude and distribution of international waters

In this chapter, the various issues associated with international waters will be discussed only in the context of fresh water. Coastal and ocean waters will not be considered, since these sources need to be analysed and discussed in a different context and within wholly different frameworks.

In the area of fresh water, three types of water sources need to be considered: rivers, lakes, and groundwaters, even though these sources are often interconnected. While we now have considerable experience with the management of international rivers, and to a lesser extent international lakes, similar knowledge on groundwater issues is conspicuous by its absence. Because of space and time constraints, the focus of this chapter is exclusively on the management of international rivers and lakes.

The global magnitude and distribution of the problem of international rivers and lakes are basically unknown at present, even though the literature is replete with unscientific and hypothetical "facts and figures". The information base for international aquifers is fundamentally non-existent.

It has generally been assumed in the past two decades that there are 214 international river and lake basins in the world. This number originated from a report completed in 1976 and published in 1978 by the now-defunct Centre for Natural Resources, Energy, and Transport (CNRET) of the Department of Economic and Social Affairs of the United Nations. This number itself was an upward revision of an earlier (1958) CNRET estimate of 166 international river and lake basins.

While the 1976 CNRET study was a useful contribution, its shortcomings are clearly evident more than two decades later. The study defined a river basin as an "area within which waters of natural origin (rain, groundwater flow, melting of snow and ice) feed a given river". It considered only those international river basins which were "separate" (that is not tributary basins), and were connected "directly with the final recipient of the water (oceans, closed inland seas and lakes)". The distribution of the number of international river and lake basins by continents, according to the CNRET study, was as shown in Table 1.1.

The study further indicated that nearly 47 per cent of the world's landmass (excluding Antarctica) falls within international basins. It ranged from a high of nearly 60 per cent of the landmass in Africa and South America to a low of about 40 per cent in North and Central America. Viewed in a different fashion, the report indicated that there

Table 1.1 Distribution of international river and lake basins by continent

Africa	57
Asia	40
Europe	48
North and Central America	33
South America	36
Total	214

are 44 countries where at least 80 per cent of the total landmass is within international basins. Of these 44 countries, 20 are in Africa, seven in Asia, 13 in Europe, and four in Latin America.

The CNRET study can at best be considered a preliminary assessment of the magnitude of the problem, which, incidentally, was also the intention of its originators. It suffered from some serious methodological shortcomings, however, and was based exclusively on maps available at the United Nations map library. It was exclusively a desk study with maps, some of which were on a scale of 1:15,000,000, or even less.

Unfortunately, the results of the CNRET study have been cited often in the past without any technical scrutiny; not surprisingly, these figures are now accepted as facts. This uncritical acceptance of the written word is all the more difficult to justify when one considers the fact that during the intervening two decades many countries, like the Soviet Union, Czechoslovakia, and Yugoslavia, have been split up into new countries, thus creating new international river and lake basins.

In retrospect, the uncritical acceptance of the number of international river and lake basins has had at least one unfortunate side-effect. Since the number is unquestionably an underestimate, it has given the erroneous impression that the overall magnitude and extent of the problem are much less serious than they actually are at present, and are likely to became in the foreseeable future. This issue has now been finally resolved by the work of Wolf *et al.* (1999).

This underestimation has been further compounded by the fact that international organizations, like the various United Nations agencies, the World Bank, and the regional development banks, for the most part have shied away from the issue of the management of international basins, except in non-controversial and non-threatening measures like expert group meetings, which in reality often achieve very little. These organizations have progressively become more and more risk-averse, a subject that will be discussed in the next section. The above two factors, to a significant extent, can explain why the issue of the management of international waters has not been as high up in the international political agenda over the past two decades as it should have been.

Role played by international organizations

During the past two decades, international organizations have played a very limited role in terms of facilitating agreements on international river basins. Unquestionably the most noteworthy and successful case where an international organization played a critical role as a catalyst to get the co-basin countries to agree to a treaty was the Indus river basin between India and Pakistan. Eugene Black, the then president of the World Bank, clearly and unambiguously indicated to the leaders of India and Pakistan at the highest political levels his own personal interest in resolving the conflict over the Indus basin amicably and speedily. He not only made the resources of the World Bank available to both countries for mediation, but also kept himself fully briefed of progress during the almost decade-long negotiations. When there was an impasse, he was not afraid to play a critical role in assisting the countries to overcome it. The World Bank played the role of an "honest broker" properly and impartially. The "carrot" that the World Bank extended in terms of offering to finance new development projects, subject to a mutually acceptable agreement between the two countries on the sharing of the waters of the Indus river system, proved to be a very attractive incentive for both parties conceded. The Indus treaty was formally signed by India and Pakistan on 19 September 1960 (Biswas 1992).

It is interesting to note that although the negotiations between the two countries took less than a decade, the subsequent agreement between the four provinces of Pakistan on the allocation of this water took an additional three-and-a-half decades to negotiate!

The entire negotiating process between the two countries took a remarkably short period for the successful completion of such a complex treaty. The Indus treaty is a major tribute to the astute and dynamic leadership of Eugene Black, who not only accepted the risk of potential failure but was also prepared to get involved personally and had no hesitation in putting his own reputation on the line.

The most unfortunate aspect during the 1960–1996 period was the almost total absence of the courageous and prudent leadership shown by Black, in either the World Bank or any other international organization. In 1976 another World Bank president, Robert McNamara, did discuss the issue of the sharing of the Ganges waters between India and Pakistan (then including Bangladesh), but the topic did not make any progress for many reasons. It was the technical professionals at the World Bank who were interested in the resolution of the problem. By the mid-1970s the Ganges issue had already been highly politicized in the countries concerned. The World Bank career professionals had very little, if any, access the highest levels of political decision-making, especially in India. Without such high-level access, it was impossible to find a solution that would have been politically acceptable to all the countries concerned. What is more, in contrast to the "honest broker" role played by the World Bank for the Indus river treaty, it was no secret that the bank's own idea of the Ganges solution was closer to the one advocated by Bangladesh. Not surprisingly, India did not trust that the bank could play an impartial role in any mediating process, and thus it summarily rejected the bank's overture.

In addition to all these factors, it has to be admitted that the World Bank's credibility with and overall impact on the developing countries were significantly higher in the 1950s as compared to the 1970s and later.

Two other international organizations have attempted to play a role in managing international rivers: the United Nations Environment Programme (UNEP) on the Zambesi river, and the United Nations Development Programme (UNDP) on the Mekong river.

The Zambesi basin covers eight countries: Angola, Botswana, Malawi, Mozambique, Namibia, Tanzania, Zambia, and Zimbabwe. UNEP convened a conference on the environmental management of the common Zambesi river system in Harare, Zimbabwe, in May 1987. The primary objective was to approve the draft Zambesi action plan (ZACPLAN), which was prepared by UNEP in close consultation with most of the countries concerned. The plenipotentiaries of five co-basin countries (Botswana, Mozambique, Tanzania, Zambia, and Zimbabwe) did sign an international agreement on the action plan for the environmentally sound management of the common Zambesi river system (David 1988; Nakayama 1997). While initially considered to be a success for UNEP, real progress in terms of its implementation of the plan in more than a decade has been minuscule.

The UNDP played a role in facilitating the agreement on cooperation for the sustainable development of the Mekong river basin. It is too early to make any judgement on the impact of this agreement. However, the fact remains that the UNDP had played "the roles of godfather, referee, rich uncle and fund-raiser to the Mekong Committee" (Miller 1996), which had already spent hundreds of millions of dollars. The disappearance of the Interim Mekong Committee, which seemed to be a distinct possibility in 1992, would have been a serious moral blow to the UNDP, since all this investment would have been lost. The UNDP subsequently played a very constructive role, which led to an agreement being signed on the lower Mekong by Cambodia, Laos, Thailand, and Viet Nam in 1995. China, the most powerful country in the region and the most upstream one on the river, was not a signatory to the agreement; it is also not a member of the new Mekong River Commission. This is likely to be a major constraint in the future for any basinwide integrated development.

While the agreement on the lower Mekong is a step in the right direction, it is unlikely to contribute significantly to the sustainable development of the Mekong river basin for the following reasons.

First, water requirements for all uses in China are increasing rapidly, and thus it needs to develop its water resources as much as possible and as quickly as possible. Currently several provinces of China have plans to develop the water resources of the upper Mekong. China has steadfastly refused to join either the earlier Interim Mekong Committee or the present Mekong River Commission. Also, the fact that China was one of the three countries which voted on 21 May 1997 against the resolution to establish an international convention on the non-navigational uses of international watercourses at the United Nations may have some negative implications for the future development of the Mekong.

Second, the most difficult part of any international agreement in the developing world is the actual water allocation between the various co-basin countries. The Mekong agreement does not include any specific allocation of water between the countries. It contains terminologies like "reasonable and equitable utilisation", and "prevention and cessation of harmful effects", on which even independent, objective experts may not agree, let alone countries with vested interests. The potential for conflicts between the countries on this issue in the future is therefore very high.

Thus, overall, no matter what criteria are used for analysis, international organizations have played a very marginal role in resolving conflicts on international rivers and lakes in recent decades. Given real leadership in the major international institutions, they could have played a significantly more effective role in this area over the past 35 years. Regrettably, there are no signs that this situation is changing for the better.

Resolution of locational water conflicts

Since water is mobile and does not respect political boundaries, countries on an international river could use its resources sequentially as it moves downstream. When a river forms a boundary between two countries, two political units have simultaneous authority over it, which could contribute to the generation of several types of transboundary water conflicts.

Because of potential conflicts between co-basin countries on international rivers and lakes, historically many nations have negotiated mutually acceptable agreements as to how such water bodies can be used. Over 3,600 treaties can be noted on international water bodies between AD 800 and 1985, the majority of which deal with navigation – water was the primary means of transportation in earlier times. Especially after the Second World War, many treaties were negotiated on international water bodies dealing with non-navigational issues like flood control, hydropower development, water quality management, and water allocation. It should be noted that generally it has been easier to negotiate treaties on the navigational uses of rivers, since these do not require water allocation or use considerations. Historically, it has been most difficult to reach agreement on the actual allocation of water quantities between the relevant co-basins countries.

The first important study on the legal aspects of using the waters of international rivers was carried out by Professor H. A. Smith of London. His book on *The Economic Use of International Rivers* was published in 1931, and reviewed more than 100 treaties and studied several conflicts on the use of international rivers. He emphasized the doctrine of riparian rights, which entitles the lower riparian states to a share of the natural flow of a river. He also noted that some of the treaties considered the concept of equitable utilization.

In 1956, the International Law Association (ILA) produced the Dubrovnik Rules for international rivers. In 1959, Bolivia introduced a resolution in the United Nations General Assembly which requested the Secretary-General to prepare a report on laws related to international rivers. This resolution was passed, although Bolivia's motivation for proposing it this is unknown.

In 1966, the ILA adopted the so-called "Helsinki Rules" for international watercourses. In 1970, Finland introduced a resolution in the UN General Assembly on the laws for international watercourses which suggested that the Helsinki Rules should be considered as a model.

During the ensuing discussions, three reservations concerning the Helsinki Rules surfaced.

- The rules were formulated by a professional organization which did not represent nation-states.
- Since nation-states had not participated in the formulation of the rules, their adoption as a model could preclude new considerations on this complex issue.
- The rules were based on a drainage basin approach, which could be a potential threat to national sovereignty.

An analysis of the ensuing discussions indicates that the most important reservation was use of the drainage basin concept. Belgium, Brazil, China, and France argued that such an approach would be a radical departure from the traditional channel-based international law. In contrast, Finland and the Netherlands felt that the drainage basin framework was the most rational and scientific approach. Certain other countries opined that the problem of international river basins was so diverse that codification may be not possible.

The resolution to refer to the Helsinki Rules as a model was lost.

However, a similar resolution was passed with only one negative vote (Brazil) after the reference to the Helsinki Rules was deleted. This resolution recommended that the International Law Commission (ILC) should "take up the study of the law of the non-navigable uses of international watercourses with a view to progressive development and codification".

In 1974, the ILC sent out a questionnaire to all members of the General Assembly soliciting their view on nine key questions. The responses, however, were not encouraging. By 1975, only 21 of the 147 UN members had bothered to reply. Four additional countries replied by 1978, one in 1979, four in 1980, and two more by 1982. Not surprisingly, on the critical issue of the appropriateness of the drainage basin concept, the countries were sharply divided. Approximately half supported the concept (Argentina, Finland, and the Netherlands) and the other half were either strongly negative (Austria, Brazil, and Spain) or ambivalent. Because of such sharp differences, the ILC decided to begin with the formulation of general principles, and then determine the scope of the term "international watercourses" later. This issue was finally addressed in 1991, when the ILC produced a draft report on the law of the non-navigational uses of international watercourses.

Considerable discussion took place on the ILC draft between 1991 and 1997. Finally, on 8 July 1997, the UN General Assembly approved the resolution on non-navigational uses of international watercourses. It is interesting to review the voting pattern on this resolution, especially in terms of existing disputes on international basins. Bangladesh, Brazil, Cambodia, Jordan, Laos, Nepal, South Africa, Sudan, Syria, Thailand, and Viet Nam were in favour; Burundi, China, and Turkey were against; and Argentina, Egypt, Ethiopia, France, India, Israel, and Pakistan abstained.

The General Assembly resolution adopted the Convention on the Non-Navigational Uses of International Watercourses, and it is now open for signature by the states. Even though some 106 countries voted for the convention, it is unlikely to be ratified in the foreseeable future because of the complexity and sensitivity of the issue.

The future

If and when a convention on non-navigational uses of international watercourses is ratified, a major issue will be its potential impact in terms of resolving existing and future disputes. In all probability it will have very marginal impact on the resolution of the conflicts, for several reasons.

Table 1.2 Current water conflicts and voting patterns on UN Convention on Non-Navigational Uses of International Watercourses

Euphrates/Tigris	Syria in favour, Turkey against, and Iraq not involved
Ganges	Bangladesh and Nepal in favour, but India abstained
Jordan	Jordan in favour but Israel abstained
Mekong	Cambodia, Laos, Thailand, and Viet Nam in favour, but China against
Nile	Sudan in favour, Egypt and Ethiopia abstained, and Burundi against
Plata	Brazil in favour but Argentina abstained

First, not all countries that are currently parties to disputes on international watercourses are likely to sign the convention. Table 1.2 shows some of the current conflicts and the voting patterns on the proposed convention in the UN General Assembly of the countries concerned.

This probably means that if and when the convention is ratified, some parties to specific conflicts are unlikely to be signatories. As past experience with the nuclear non-proliferation treaty has shown, moral pressures are likely to be of little value when faced with strong, entrenched, vested national interests; experience with this new convention is unlikely to be any different.

Second, while the 1997 convention could be considered to be an important benchmark, its two basic principles are similar to what had generally been accepted much earlier: equitable and reasonable utilization and an obligation not to cause appreciable harm. Thus the proposed convention did not break any new ground, at least conceptually.

One of the main problems with the proposed convention is that it is full of vague, broad, and general terms (Waterbury 1997) which can be defined, and in certain cases quantified, in a variety of different ways. Accordingly, expert advice can be easily "tailored" to legitimize each country's political views and demands. Technical analyses can be produced to justify and support appropriate national positions. Such situations, however, are not new: they have happened in the past and will no doubt continue to occur in the future. Furthermore, the convention does not give any practical guidance to the negotiators, nor any operational assistance to the technical experts. It simply outlines a very broad, general framework, within which everything is considered to be relevant and important. It is likely to contribute to the generation of significant differences of opinion among the negotiators and technical experts as to how each general article should be interpreted in operational terms.

One can argue that the convention outlines certain factors which could determine one of the fundamental principles, that of "equitable and reasonable use". According to the convention, such a process should take "into account all relevant factors, and circumstances", including:

- geographic, hydrographic, climatic, ecological, and other factors of a natural character;
- social and economic needs of co-basin countries;
- effect of the uses of the watercourse on other co-basin states;
- existing and potential uses of the watercourse;
- conservation, protection, development, and economy of use of the watercourse resources and the cost of measures taken to that effect;
- availability of alternatives, of corresponding value, to a particular planned or existing use.

None of these factors can be defined uniquely or precisely, since they are general and broad in character. Accordingly, when all the factors are integrated to define "equitable and reasonable use", the countries in conflict would find it a very difficult task to arrive at mutually acceptable estimates, and estimates are likely to differ significantly even when groups of truly independent and objective experts make such attempts separately.

Third, the prevailing national political sentiments in each negotiating riparian country, as well as the incentives to negotiate in good faith, are likely to be important factors in the resolution of all such conflicts. In addition, as the number of riparians increase to four or more in any international basin, the importance and relevance of any proposed settlement could range from exceedingly important for one country to total indifference or even downright hostility from another. For a basin like the Nile, which has 10 riparians countries, the incentives for all parties to arrive at any specific settlement are likely to range from very high to of no discernible interest. Equally, the types of settlements preferred by the different countries are likely to vary significantly. The new convention can at best be of limited help in such cases.

Finally, ratification of the convention is an important requirement. The convention can enter into force on the "ninetieth day following the date of deposit of the thirty-fifth instrument of ratification, acceptance, approval or accession with the Secretary-General of the United Nations". The convention is open for signature until 20 May 2000.

Nearly two years after the approval of the convention by the General Assembly, only four states (Finland, Norway, South Africa, and the Syrian Arab Republic) have so far ratified it. Whether another 31 states will ratify it before the closing date is an open question. Accordingly, the potential impacts of the convention on the resolution of disputes over international rivers and lakes cannot be surmised at present.

The above considerations and other related factors mean that agreements in individual international basins will continue to occur only through protracted negotiations between the riparians concerned. The convention, even if it is ratified, is unlikely to speed up the time needed to reach mutually acceptable agreements in the vast majority of disputes on international river basins.

Conclusion

As the twenty-first century dawns, it is becoming evident that, like oil some two decades ago, the era when water could be considered to be a cheap and plentiful resource is now virtually over. Increasing water demands, limited availability of this resource, and higher levels of contamination mean that the water management profession will face a problem of a magnitude and complexity which no earlier generation has had to face. The water profession really has two fundamental choices: to carry on as before with only incremental changes and a "business-as-usual" attitude, and thus endow future generations with a legacy of inefficient water management practices and potential serious conflicts on international water bodies; or to continue in earnest with an accelerated effort to plan, manage, and use international watercourses sustainably and fairly.

The root for the English word "*rival*" is from the Latin term *rivals*, which originally meant using the same river (*rives*). But as the world becomes increasingly more and more interconnected, countries sharing the same river should no longer consider each other as rivals. With properly conceived frameworks, management and use of international water bodies should result in "win-win" situation for all the parties concerned. Contrary to popular belief, these are not zero-sum games.

REFERENCES

Biswas, Asit K. 1992. "Indus Water Treaty: The Negotiating Process", *Water International*, Vol. 17, pp. 201–209.

Biswas, Asit K., N. V. Cordeiro, B. P. F. Braga, and C. Tortajada-Quiroz. 1999. *Management of Major Latin American Rivers: Amazon, La Plata and Sao Francisco*. Tokyo: United Nations University Press.

Biswas, Asit K. and T. Hashimoto, eds. 1996. *Asian International Waters: From Ganges-Brahmaputra to Mekong*. New Delhi: Oxford University Press.

Centre for Natural Resources, Energy and Transport. 1978. *Register of International Rivers*. Oxford: Pergamon Press.

David, L. J. 1988. "Environmentally Sound Management of the Zambesi River Basin", *International Journal of Water Resources Development*, Vol. 4, No. 2, pp. 80–102.

International Irrigation Management Institute (IIMI). 1992. *Developing Envi-*

ronmentally Sound and Lasting Improvements in Irrigation Management: The Role of International Research. Colombo: IIMI.

Miller, M. 1996. "Transformation of a River Basin Authority: The Case Study of Mekong Committee", in *Asian International Waters: From Ganges-Brahmaputra to Mekong*, eds A. K. Biswas and T. Hashimoto. New Delhi: Oxford University Press, pp. 226–244.

Nakayama, M. 1997. "Successes and Failures of International Organisations Dealing with International Waters", *International Journal of Water Resources Development*, Vol. 13, No. 3, pp. 367–382.

United Nations, Report of the Secretary-General. 1997. *Comprehensive Assessments of the Freshwater Resources of the World*. Committee on Sustainable Development, Fifth Session, E/CN. 17/1997/9. New York: United Nations.

Waterbury, J. 1997. "Between Unilateralism and Comprehensive Accords: Modest Steps toward Cooperation in International River Basin", *International Journal of Water Resources Development*, Vol. 13, No. 3, pp 279–290.

Wolf, A. T. *et al.* 1999. "International River Basins of the World", *International Journal of Water Resources Development*, Vol. 15, No. 4, pp. 387–428.

World Bank. 1992. *World Development Report*. New York: Oxford University Press.

2

Ganges-Brahmaputra: The outlook for the twenty-first century

R. B. Shah

Introduction

The Ganges-Brahmaputra-Meghna (GBM) river basins, encompassing Nepal, India, Bangladesh, the autonomous Tibet region of China, and Bhutan, have considerable water resources potential which can be harnessed for multipurpose beneficial uses – like irrigation, hydropower generation, flood control, navigation, fishery development, etc. – in the countries forming a part of these mighty river basins through cooperation among those countries in order to make optimal use of this potential for common good. Incidentally, this region constitutes one of the poorest regions in the world. Global experience clearly shows that development of water resources ushers in an era of all-round prosperity by increasing agricultural production through irrigation, by industrial development with the increased availability of electric power, and by preventing/mitigating the recurring damage caused by frequent floods. These trigger a chain of economic activities with forward and backward linkages, which increase employment potential considerably and result in an improvement in the standard of living of the people, and in turn of the nations involved.

However, so far very little of this precious natural gift has been utilized in these river basins. Consequently, year after year, the waters of these river systems continue to flow unused to the seas, causing flood damage en route in most years.

Meanwhile, with ever-increasing population growth and increasing demands on water for various competing uses, available water is becoming scarce, the cost of development is escalating, misconceived opposition to the conservation and development of water resources is growing, and, at the same time, the degradation of not only the environment but also the quality of life of the people of this region is continuing at a faster rate.

At the end of the twentieth century, one gets an inescapable feeling of sadness in having lost/wasted the century as far as the development of water resources and their utilization in this region are concerned.

This chapter, therefore, is a modest effort to highlight the enormous potential, make an honest assessment of the efforts and limited achievements so far, and summarize the outlook needed and the tasks ahead for the twenty-first century.

Basic facts

To have a proper appreciation of the problem and the perspective, it is necessary to note the basic indisputable facts which constitute the parameters for any growth model. While the Ganges-Brahmaputra and the Meghna are three distinct river basins, these merge into one as far as Bangladesh is concerned. As they are adjoining basins with some degrees of commonality as well as disparity, it is desirable to study them together as one entity, as there is a scope of complementality. Therefore, this chapter covers the region traversed by the Ganges, Brahmaputra, and the Meghna. In dealing with details, the Tibet region is left out, as many facts and figures concerning it are not available and the impact of developments there may not be significant in other parts of the region. Some of the basic facts in the region pertaining to these river systems are given in the following sections.

Geographical area

The total area traversed by these rivers is around 1.65 million square kilometres. Distribution of geographical area in the countries of the basins is given in Table 2.1. The percentage of the total geographical area of the GBM region contributed by various countries (Char 1997a) is as given in Table 2.2.

Population

The total population in the GBM region is about 536 million. The breakdown by country is given in Table 2.3. The GBM region constitutes about 0.12 per cent of the world's total geographical area, but it has to

Table 2.1 Geographical area of the GBM region by country

	Basins (million ha)				
	Ganges	Brahmaputra	Meghna	Minor rivers	Total
India	86.14	19.44	4.17	3.63	113.38
Nepal	14.70	–	–	–	14.70
Bhutan	–	4.50	–	–	4.50
Bangladesh	3.90	4.70	3.62	–	12.22
Total	104.74	28.64	7.79	3.63	144.80

Table 2.2 Percentage of GBM geographical area by country

Country	Ganges %	Brahmaputra %	Meghna and minor rivers %	GBM %
India	82.24	67.88	68.30	78.30
Nepal	14.03	–	–	10.15
Bhutan	–	15.71	–	3.11
Bangladesh	3.73	16.41	31.70	8.44
Total	100.00	100.00	100.00	100.00

Table 2.3 Regional population

Country	Population (millions)	% of total GBM population	Population density/ha
India	405.4	75.66	3.57
Nepal	18.9	3.53	1.28
Bhutan	1.5	0.28	0.33
Bangladesh	110.0	20.53	9.00
Total	535.8	100.00	3.70

support about 10 per cent of the world's total population. Most of the population, varying from about 70 per cent in India to 90 per cent in Nepal, works in agricultural-based employment.

Poverty

According to an assessment by the World Bank, about half of the world's poor live in South Asia. Within the South Asian region, the Ganges-Brahmaputra-Meghna region contains the largest concentration of the poorest in the world, representing about 40 per cent of the total number of the poor in the developing world.

Table 2.4 Human deprivation parameters in the GBM region

Parameter	India %	Nepal %	Bangladesh %	Bhutan %
Population in absolute poverty	46	40	52	N/A
Population without access to safe water	25	55	22	66
Population without access to sanitation	71	79	66	87
Human development index	0.436	0.332	0.365	0.307
GNP per capita (US$)	300	190	220	170

Table 2.5 Average annual rainfall in GBM countries

Country	mm
Nepal	1,520
India (whole country)	1,300
Bangladesh	2,400
Bhutan (higher altitudes)	500
Bhutan (southern plains)	2,000
Bhutan (foothills)	5,000

There are various parameters/indicators of poverty. Prasad (1997) states that the UNDP estimated poverty index for Bangladesh is 86.5 per cent, for Nepal 60 per cent, and for India 55 per cent. In 1993, some of the human deprivation profile parameters were as given in Table 2.4 (Rangachari 1997).

The figures in Table 2.4 are more-or-less representative in respect of Nepal, Bhutan, and Bangladesh. However, in the case of India, national averages are not valid for the GBM region. States of India like Assam, Bihar, Madhya Pradesh, Rajasthan, Uttar Pradesh, and West Bengal, which form a part of this region, are below the national parameters and constitute a major poverty subregion in India.

Rainfall

The average annual rainfall in some countries of this region is given in Table 2.5.

Most of the annual rainfall (75–80 per cent) occurs in the three to four months of the south-west monsoon. Even in this period, the bulk of the rainfall comes in three to five spells of a few days each. Thus, most of the water is received in a period of 30–40 days, but it has to last for the entire year. There are also variations in the quantum of rainfall from year to

Table 2.6 Average annual surface water flows in GBM countries

Country	Million ha-m Ganges	Brahmaputra	Meghna	Total
India	52.50	53.72	7.93	114.15
Nepal	20.00	–	–	20.00
Bangladesh	3.40	10.25	5.10	18.75
Bhutan	–	N/A	–	–
Total	75.90	63.97	13.03	152.90

Table 2.7 Available groundwater resources

Country	Million ha-m Ganges	Brahmaputra	Meghna	Total
India	17.17	2.78	0.18	20.13
Nepal	1.20			1.20
Bangladesh	N/A	N/A	N/A	3.22
Bhutan	–	N/A	–	N/A
Total	–	–	–	24.55

year, apart from spatial and temporal variations. Bangladesh is subject to frequent cyclonic storms bringing intense spells of rainfall.

Available water resources

Surface water

On the basis of available published figures, average annual surface flows in various river basins in different countries are given in Table 2.6. In terms of water resources availability, the GBM is the second largest river basin in the world, after the Amazon river basin.

It will be seen from Table 2.6 that India contributes about 75 per cent of the water resources of the entire GBM region – comprising 69 per cent of the Ganges basin, 84 per cent of the Brahmaputra basin, and about 61 per cent of the Meghna and minor rivers. In fact, out of the total surface water resources in India, about 63 per cent are in these river basins.

Groundwater

The available groundwater resources are roughly as given in Table 2.7. Groundwater resources as per present estimates are about 16 per cent of the available surface water resources.

Utilizable water resources

Surface water

On account of the fact that bulk of the surface water resources are derived from south-west monsoon rainfall, falling over a total period of 30–40 days per year in short spells spread over the monsoon season, unless these waters are stored in minor, medium-sized, and large reservoirs for subsequent multipurpose use throughout the year, they cannot be utilized.

Therefore, ultimate utilizable surface water resources in India have been estimated as 32 per cent of the available resources of the Ganges with dams in India, or 47 per cent of the Ganges with dams in India and Nepal, and 4 per cent of the Brahmaputra and Meghna.

Thus, the maximum utilizable surface water resources in India in the GBM region will be about 19 million ha-m, which may rise to about 27 million ha-m with the construction of dams in Nepal. Since all the possible dams may not materialize, the utilizable surface water resources can be considered as about 20–24 million ha-m.

In the case of the Ganges basin, with the limited availability of suitable sites for construction of storage dams in India, utilizable water in India is about 32 per cent. Since most of the important tributaries of the Ganges rise in Nepal before flowing down to India, and storage sites on these tributaries are available in Nepal, the utilizable quantum of Ganges water in India will increase to the extent that storage dams are built in Nepal. Even so, more than 50 per cent of Ganges water from India will continue to flow unused to the sea, mainly in the monsoons.

Similarly, it is quite evident that only 4 per cent of available water in the Brahmaputra-Meghna basin can be utilized in India, and the balance will continue to flow unused to the sea through Bangladesh; even this minimal utilization will only be realized provided a few dams which are feasible within Indian territory are actually constructed.

Similar estimates for other countries are not available. Since the consumption requirements for water in Nepal are not likely to be very high, and since there is no possibility of constructing dams in Bangladesh, total utilizable surface water resources can be estimated as about 20–24 million ha-m, which is only about 13–15 per cent of available water resources. This constitutes one of the major constraints in development of water resources in this region.

Groundwater

As far as groundwater is concerned, the entire available potential again cannot be utilized. Normally, 80 per cent of the annual recharge is utilized so as not to affect the groundwater dynamic equilibrium adversely.

In India, in some states, adverse effects of overexploitation of groundwater in terms of continuous decline in water levels, appearance of harmful chemicals like nitrates, arsenic, etc., and intrusion of salinity in coastal areas are already being experienced. Thus, about 20 million ha-m of groundwater may be considered available for utilization. Most of the rural water supply for drinking and domestic use is from groundwater. In addition, in the Ganges basin in India, groundwater is being extensively used for irrigation.

Therefore, for future development, groundwater will not be a major source, and future planning for development and utilization of water resources has to take into account these facts/constraints.

Hydropower

The GBM region, with its rivers originating in Himalayan and sub-Himalayan ranges with steep topographical slopes and almost year-round availability of flows, has a tremendous hydropower potential. Although there are varying estimates of the potential, generally accepted figures are that India has 45,635 MW (economically exploitable), Nepal has 83,000 MW (estimated) and 42,000 MW (economically exploitable), and Bhutan has 21,000 MW.

The potential developed so far and under creation is indeed meagre, as may be seen from the figures of installed capacity in India of 3,525 MW, Nepal 244 MW, Bhutan 336 MW, and Bangladesh 230 MW. Thus, the hydropower potential developed/under development in the entire GBM region is only 4,335 MW, which is about 5 per cent of economically exploitable potential.

This brings out the tremendous scope for future development, which will not only help in boosting agricultural and industrial development in the region as a whole, but will particularly help in considerable economic development of Nepal and Bhutan, as hydropower is the most precious natural resource and asset for these countries and perhaps the only major resource for their economic growth. For both of these countries, India, where the supply-demand gap is ever increasing, will be the market.

Flood control

While about 60 per cent of India's flood-affected area lies in the Ganges-Brahmaputra-Meghna basins, almost 80 per cent of the entire area of Bangladesh is prone to floods. The areas subjected to floods are the least in Bhutan, as the rivers flow in deep incised channels. The problem of floods in Nepal is also comparatively limited for the same reason, although areas in the inner terai and terai regions are prone to flooding. The floods are caused by incessant rains, which also contribute to landslides.

It is universally recognized that the only method of really controlling the floods is the detention of flood waters in storage reservoirs and subsequently releasing the water in a regulated fashion. There are no storage reservoirs constructed so far in India in the Brahmaputra and Meghna basins. On the northern tributaries of the Ganges there are hardly any adequate storage sites available in India, although a number of sites exist in Nepal. However, so far no major dams have been constructed in Nepal. In India, some dams on the southern tributaries of the Ganga have been constructed, but, except for a dam on the Ramganga, no major dam on the northern tributaries has been completed. Another major dam, Tehri, is under construction. There is no scope for constructing any dams in Bangladesh, which is essentially a deltaic region.

Because of this, major portions of the GBM region in India and Bangladesh are subjected to recurring flood damage, and will continue to be so to an appreciable extent.

A brief review of past cooperation

India-Nepal cooperation

The earliest attempt at Indo-Nepal cooperation in the field of water resources dates back to 1920–21, when, under the British rule, an agreement was reached on the Sarada barrage on the river Mahakali in India. This envisaged the diversion of 460 cusecs (flow of cubic feet per second) of water during the wet season and 150 cusecs during the dry season to Nepal through an independent canal; meanwhile, the Sarada barrage has provided irrigation over a large area in the Indian state of Uttar Pradesh.

In the post-independence period, cooperation between India and Nepal began with the implementation of the Kosi barrage at Bhimnagar and the Gandak barrage at Baisaltan, both financed by India under agreements signed for the two projects in April 1954 and December 1955 respectively; both the agreements were subsequently revised in 1966. The barrages provide benefits of irrigation, hydropower, and flood control (to some extent) to both countries. Under these two projects, some irrigation benefits are extended to Nepal as well as the supply of hydro-electric power. Since no storage dams have been constructed on these rivers upstream, the existing irrigation benefits are basically from run-of-the-river operation.

India has also assisted Nepal with both technical and financial assistance in building hydro-electric projects, such as Trisuli (21 MW), Kataiya (20 MW), Gandak (15 MW), and Devighat (14 MW). In addition, India provided assistance for the modernization of Chatra and Chandra canals for stabilizing and extending irrigation benefits in Nepal.

Although a number of other major multipurpose water resources projects on the important tributaries of the Ganges like the Kosi, Rapti, Sarada, Gandak, Bagmati, etc. have been under formulation and discussion for a number of years, no tangible results have been achieved so far.

However, on 12 February 1996 a treaty between the governments of India and Nepal was signed concerning the integrated development of the Mahakali river, including the Sarada barrage, Tanakpur barrage, and the Pancheswar project. This was a significant benchmark in cooperation after the unfruitful results of discussions at various levels over the preceding 30 to 40 years. This treaty was ratified by both the countries and instruments of ratification exchanged during the visit of the Indian prime minister to Nepal in June 1997. The multipurpose Pancheswar project is envisaged as having an installed capacity of 2,000 MW/6,480 MW of hydropower and irrigation and flood control benefits to both countries; the agreed detailed project report, including the installed capacity for hydropower, is yet to be finalized.

During the Indian prime minister's visit in June 1997, a power trade agreement was also signed.

India-Bangladesh cooperation

Following the visit of the prime minister of India to Bangladesh in March 1972, the Indo-Bangladesh Joint Rivers Commission (JRC) was set up in November 1972 for harnessing the rivers of the two countries for the benefit of both. Initially, the commission was entrusted with matters related to flood management and the resolution of common border river problems. Subsequently, in 1974, the JRC was entrusted with the additional task of resolving the issue of sharing flows of all the common rivers, and the formulation of agreed proposals for augmenting the flows of the Ganges at Farakka.

Interim arrangements for sharing lean flows of the Ganges at Farakka were formulated for short periods during 1977, 1982, and 1985. The last interim sharing arrangement expired on 31 May 1988; because of the failure of both the countries to arrive at any amicable solution for augmenting the flows at Farakka, there was no agreement for sharing lean flows of the Ganges after 1988. However, on 12 December 1996 a new treaty for sharing Ganges waters at Farakka for the lean season was signed. In this treaty, the issue of augmentation of lean-season flows at

Farakka has not been made contingent to the sharing arrangement. However, the treaty does stipulate the need to cooperate with each other in finding a solution to the long-term problem of augmenting the flows of the Ganges during the dry season. This treaty will remain valid for a period of 30 years and shall be renewable by mutual consent. This has been a significant landmark in Indo-Bangladesh cooperation, as it removes one of the major irritants.

The issue of lean-season sharing of the waters of the Teesta river and 54 other small rivers common to both countries is still under consideration.

In the meantime, the government of Bangladesh has revived its proposal to construct a barrage on the Ganges at Pangsha, about 60 kilometres downstream of the Hardinge bridge in Bangladesh. This project will be multipurpose in character; India has welcomed the proposal and has offered technical assistance for preparation of detailed feasibility reports. It is understood that the government of Japan is likely to provide financial assistance for investigations and the preparation of a feasibility report for this project.

India-Bhutan cooperation

Bhutan has a hydropower potential in the order of 20,000 MW. India has been giving both technical and financial assistance to Bhutan for about the last 40 years for various development activities, including development of hydropower. Since the commissioning of the first micro-hydro-electric project in Thimpu (400 kW) in 1966, a number of other micro-hydro-electric stations have been planned and built in various parts of Bhutan, with a total installed capacity of about 10.5 MW. The first major hydro-electric project, Chukha, with 336 MW installed capacity, was constructed under an agreement signed in 1975 with financial assistance from India (60 per cent grant and 40 per cent soft loan), and was commissioned in 1987. Since the requirements for power in Bhutan are extremely small, the bulk of the power from this project is exported to India under an agreed tariff which is revised at fixed intervals. With the encouraging experience of this project in terms of the economic development of Bhutan, another project, Kurichu in western Bhutan, with an installed capacity of 45 MW, is also under implementation. Another major project, Tala, second in the cascade development of the Wangchu river, with an installed capacity of 1,020 MW, has also been taken up for implementation from 1997 with financial assistance from India (60 per cent grant and 40 per cent loan). Both the Chukha and Tala projects are being implemented through project authorities specially set up; these are autonomous, with high-level representation from both India and Bhutan, and are chaired by a representative of the king of Bhutan.

There are several other major hydro-electric projects in various stages of planning, investigation, and preparation of project reports, such as Wangchu Stage II (900 MW), Sankosh (4,060 MW), and Bunakha (180 MW) (Char 1997).

Indo-Bhutan cooperation is very significant in terms of economic development in Bhutan, where the revenue earned from hydropower development will be the main source of income for the country to help in its overall economic development and raising standards of living. At the same time, it will enable India to meet its increasing demand for power.

Future areas of cooperation

Against the background of the basic indisputable facts mentioned earlier, as well as in view of the large potential for development of water resources for the overall benefit of the region, it will be obvious that there are many areas of mutual cooperation. These can be briefly summarized as:

- hydropower generation and inter-country transfer of power;
- flood control and forecasting;
- technical assistance in formulation of water resources projects;
- joint implementation and operation of major multipurpose projects;
- management and equitable sharing of lean flows;
- inland navigation;
- augmentation of flows.

While these are general, broad areas of cooperation, the forms and modalities of cooperation in each specific area will differ from one country to another. However, it needs to be emphasized that past mindsets have to be changed, and flexible rather than rigid approaches need to be adopted, with the sole aim of mutual cooperation for the benefit of the countries of the region. The old maxim "where there is a will, there is a way" is all the more applicable now. The last few years have witnessed the rapid demolition of political, mental, and attitudinal barriers the world over. Interdependence among nations is an inevitable fact of the new world order. Regional cooperation in countries geographically close to each other with many bonds of past history and similarities in culture and way of life should provide a favourable platform to foster cooperation for the common good of the poorest people inhabiting this region. Gradually, enlightened public opinion in all the countries of this region is veering round to this approach. This now needs to be crystallized in specific cooperative ventures. A broad delineation of some areas of this cooperation has been attempted in the following sections.

Hydropower generation and interconnection

Potential and market

As already mentioned, Nepal has an economically exploitable hydropower potential of 42,000 MW and Bhutan has about 20,000 MW. The market for absorption of this power is basically in India, as the requirements of Nepal and Bhutan, even on a long-term basis, will be quite small. However, transmission and absorption of the power generated in Nepal and Bhutan will have to be through an interconnected integrated power system between the various countries.

International experience

Interconnection of power systems of contiguously located countries and their coordinated operation provide immense technical and economic benefits (Sambamurty 1997). In Europe, such integrated power systems have been in existence and operation for over 70 years, covering most of the European countries like Germany, France, Italy, Belgium, the Netherlands, and Switzerland as well as the former Yugoslavia, Greece, Spain, and Portugal. Similarly, there is another integrated system between Norway, Sweden, and Finland. The USA and Canada are also sharing hydropower in the St Lawrence and Columbia river basins.

Recently, development of interconnected power systems has also commenced in the countries of Egypt, Libya, Tunisia, Syria, Morocco, Jordan, Lebanon, Iraq, and Turkey. Similarly, in Central America, Mexico, Columbia, and Venezuela have started development of interconnections between their systems. There is thus plentiful international experience available in the successful interconnection and operation of inter-country power systems. This can usefully be extended to the GBM region, where the present status is as follows.

India-Bhutan

The power generated at the 336 MW Chukha project in Bhutan is transmitted thorough 220 kV lines to Birpara in India; this provides the link between the Bhutanese power system and the eastern regional power grid of India, and enables transmission of the surplus Chukha power to India. The operation of the Chukha plant is coordinated in the eastern regional grid by the eastern regional load despatch centre in Calcutta. Similar arrangements between India and Bhutan will be extended for transmission of surplus power from the proposed 1,000 MW Tala project downstream of Chukha.

The power to be generated from other hydro-electric projects in Bhutan, like Kurichu, Chukha Stage II, etc., will be similarly interconnected to the Indian system.

India-Bangladesh

Studies of the interconnection of the Bangladeshi and Indian power systems were initiated by the Power Grid Corporation of India in 1994. Studies carried out by the two countries, discussions between them, and the interest shown by the Asian Development Bank led to the formulation of an ADB regional technical assistance project known as the India-Bangladesh Electricity Exchange Project. Under this project, the techno-economics of two 132 kV double-circuit lines to interconnect the Indian and Bangladeshi systems will be examined to gauge the feasibility of the import and export of power from India to Bangladesh and vice versa.

India-Nepal

At present, India supplies power from the Kosi and Gandak projects to the border towns and villages in Nepal. However, there is no major system of significant interconnections between the Nepalese and Indian power systems.

Nepal has already started construction of some hydropower projects, like Kali-Gandaki (144 MW). In addition, some projects are also being taken up in the private sector, like Bhote Koshi (36 MW). Nepal's electricity demand by the year 2025 is projected to be about 1,700 MW, thus the power to be generated from major hydro projects like Pancheswar will be surplus to its requirements. Such surplus power will have to be transferred to India thorough an interconnecting system. In future, major multipurpose projects having a predominant hydropower component, like Chisapani on the Karnali (10,800 MW) and Kosi (7,000 MW), are likely to come up – recently it was reported that some private foreign enterpreneurs have already shown interest in developing the Chisapani project.

A programme for promoting interconnections between the Nepalese and Indian power systems will therefore have to be planned and implemented. It may begin with interconnections across the border at 132 kV level, which will later need to be extended and strengthened when the other hydro projects in Nepal are developed.

Necessity of interconnection studies

In India, the bulk of the power potential lies in the northern and north-eastern portions of the Himalayas; even at present, the north-eastern and eastern regions are somewhat surplus in power (Sambamurty 1997).

Therefore, planning of an all-India grid is already under way. The long-term national power plan prepared by the Central Electricity Authority (CEA) in March 1997 indicated the core transmission system required to transfer power from the north-eastern to the eastern and northern regions, considering both EHV AC and HV DC alternatives. These studies will have to be supplemented as and when the possibilities of power exchanges between the GBM countries materialize. The transmission lines envisaged by the CEA, considering only the projects to be developed in India, will have several nodal substations and it may be possible to use them for interconnections.

Therefore, on the lines of the study already undertaken with ADB assistance for interconnection between India and Bangladesh, similar studies for interconnection between India and Nepal need to be undertaken to decide various techno-economic details. This will involve the compatibility of the system sizes and parameters in different countries, as well as setting up of load despatch centres and system control facilities which at present are not available. This area of cooperation, which is more or less non-controversial, needs to be pursued vigorously to plan the feasibility of interconnecting systems and work out broad techo-economic details.

Tariffs

So far, there are no problems of any type between India and Bhutan. However, in the case of Nepal the question of tariffs will have to be amicably settled. At present, it is understood that there are some basic differences over even the tariff to be charged for the surplus power from the proposed Pancheswar project – this is understood to be one of the main reasons for the non-finalization of a mutually acceptable project report. However, it will be in the interest of both the countries to resolve this issue amicably as early as possible and to lay down agreed principles for the fixing of tariffs, as the actual tariff may vary slightly from project to project.

Flood control

Extent of problem

In India, the Ganges and Brahmaputra basins are subject to recurring floods of varying intensities. It is estimated that about 55 per cent of the total flood damage in India takes place in this region. Even within the region, the intensity and recurrence of floods is higher in the Brahmaputra basin, where flooding is almost annual, compared to the Ganges basin.

Floods in Bangladesh are also a recurring phenomenon, with about 80 per cent of the country's geographical area being prone to floods. Every year, almost one-third of the area gets flooded. In the recent past, the floods experienced in 1988 were the worst in living memory: over 60 per cent of the total land area was flooded and about 50 per cent of the population was directly affected.

Floods in Nepal are comparatively less severe and restricted mainly to inner terai and terai areas.

Flood control measures so far

As already pointed out, one of the recognized methods of direct control of floods is to construct storage reservoirs for flood waters. These are more effective if located near the damage centres. In the Brahmaputra-Meghna basins in India, so far no storage reservoir has been constructed. Even in the Ganges basin, where the flood problem is more severe on its northern tributaries, no storage reservoir has been constructed excepting on the Ramganga river, a tributary of the Ganges where the flood problem is not serious.

In Bangladesh, on account of its deltaic region, there is no possibility of constructing any storage reservoirs. In Nepal there are adequate storage sites on the main tributaries of the Ganges, but no storage reservoir has yet been constructed.

Therefore, so far recourse has been taken to reduce the flood damage, particularly in urban and agricultural areas, by construction of flood embankments. In India, in the Ganges basin, about 5,800 km of embankments have been constructed, and in the Brahmaputra basin about 4,800 km of embankments, giving a total of more than 10,500 km. Similarly, it is reported that about 8,300 km of embankments have been constructed in Bangladesh.

Although the embankments have afforded some degree of protection, particularly in less severe floods, they are subject to breaches during high floods and some of the embankments also get overtopped. This is mainly on account of the fact that embankments for protection of agricultural areas are generally designed for a flood of 25-year frequency. Because of poor maintenance, the embankments are subject to severe damage during floods, and damage through breaching and overtopping is much more serious than that caused by normal floods. Embankments also adversely affect the natural drainage of the protected areas. Besides, in the case of long embankments on both banks, there is an appreciable rise in flood levels for the same magnitude of floods. In many cases, the heights of the embankments have to be increased, or they have to be protected by costly anti-erosion measures.

Due to these problems, embankments along with river training works are now generally restricted to protection of urban areas. In India, many towns and cities, including Delhi, are protected by embankments.

Prospects of storage reservoirs

On the basis of the study of all available topographical sites in India, it has been estimated that the total storage capacity which can be constructed in the Ganges basin is only 5.44 million ha-m; while in the Brahmaputra-Meghna basins it is 4.76 million ha-m. Thus, together, these reservoirs will constitute only one-third of the total storage capacity that is proposed to be created in India. The storage capacity already created in the Ganges basin is 3.22 million ha-m. On the Brahmaputra, no storage capacity has so far been created. On the Barak river in India, the Tipaimukh multipurpose project envisages a 163-m-high dam with a specific flood control storage capacity. This will control the floods in the Barak valley as well as affording some relief in Bangladesh. The project will also have a hydro-electric generating capacity of 1,500 MW, and will irrigate almost 100,000 hectares of land in India, besides augmenting lean flows which will also be beneficial to Bangladesh.

On the Brahmaputra, two major storage facilities were planned at Dehang and the Subansiri. Apart from the proposed installed hydropower capacity of 20,000 MW at Dehang and 4,800 MW at Subansiri, these two reservoirs were expected to reduce the flood levels in the main Brahmaputra downstream by about one metre. However, strong reservations due to the proposed submergence of important townships mean that three dams instead of one dam on each of these rivers are now under investigation. It is not definite whether ultimately these dams will materialize, and to what extent they will afford flood protection. In any case, since the dams will take many years to construct, the problem of floods in the Brahmaputra basin will continue as before. The same situation will more or less hold good for the Ganges basin unless some major storage dams on the important tributaries like the Kosi, the Rapti, and the Bagmati are constructed in Nepal. However, these also appear to be a distant prospect.

Thus, there is no possibility of any reliable flood control in the Ganges and Brahmaputra basins, even on a long-term basis. The focus for flood management, both in India and Bangladesh, will therefore have to be shifted to "living with floods" by reducing the loss of human lives and misery to the people and livestock. The measures to be taken to achieve this purpose are flood-plain regulation, scientific and extensive flood forecasting and flood warning systems, raising of houses, important communications and tubewells above the flood levels, and storage of adequate food and fodder on raised platforms, etc.

Table 2.8 Analysis of annual rainfall

Decade	Average mm	Maximum mm	Minimum mm
1871–1880	1,059	1,230	914
1881–1890	903	1,158	1,060
1891–1900	1,094	1,336	811
1901–1910	1,040	1,156	920
1911–1920	1,077	1,349	858
1921–1930	1,074	1,125	1,032
1931–1940	1,123	1,280	998
1941–1950	1,140	1,216	957
1951–1960	1,123	1,334	972
1961–1970	1,069	1,314	872
1971–1980	1,091	1,199	848
1981–1990	1,070	1,244	965

Source: *Theoretical and Applied Climatology*, 1994

Are floods on the increase?

There is a misconception in the public mind that the floods in this region have been on the increase in recent years due to changes in the rainfall pattern and deforestation. An exhaustive study carried out by the author (Shah 1996) indicates that the floods are not on the increase, nor are the causes attributed to the floods scientifically valid.

Analysis of the annual rainfall in India carried out for a period of 120 years from 1871 to 1990 clearly indicates that there is no trend whatsoever in the rainfall pattern. The position in respect of average, maximum, and minimum annual rainfall for each decade indicates that it has been more or less the same over the last 120 years; an analysis is given in Table 2.8.

Similarly, data on heavy storms from 1901 to 1990 were collected and analysed, as the floods occur basically from the intense rainfall during heavy storms. This analysis also indicates that over the years there has been no change in the intensity and magnitude of rainfall in heavy storms. An analysis of the most severe storms published by the India Meteorological Department also indicates that there has been no change in the one-day or three-day intensities of rainfall over a period of about 90 years.

It is expected that the same position should be valid for the rainfall pattern and intensities in Nepal and Bangladesh.

Therefore, while it can be concluded that there has been no increase in the physical magnitude of floods, there has been an increase in the resulting damage, mainly on account of the encroachment in the flood plains of various rivers, as well as an increase in economic activities in the flood plains due to the increasing pressure of population.

A popular misconception that forests reduce the magnitude of floods by absorbing a portion of the heavy rainfall is also incorrect. The studies carried out indicate that while the interception for small storms with rainfall of 1–2 mm is 80 per cent, it reduces to only 4 per cent for storms where rainfall is more than 60 mm. The major storms in India which cause floods have an intensity of 360 mm and above in one day, and interception by the forests or vegetal cover is almost negligible. Major floods are also almost invariably associated with long periods of intense rainfall, during which the soil becomes nearly saturated.

Leopold and Maddock (1961) clearly state that:

> No unequivocal or unqualified statement can be made that floods are more severe and larger in number in the present generation than those which occurred prior to settlement.

Agarwal, the reputed environmental crusader in India, states (Agarwal 1991) that:

> The Indus, Ganga and Brahmaputra valleys have always been flood prone. Even when the Himalayan mountains were uninhabited and the forest cover intact, major floods visited these valleys and disrupted human civilisation. Ancient Indian literature is full of references to devastating floods ... Environmentalists should not enter into a game of one-upmanship with water engineers. There is no evidence to believe that ecological solutions like afforestation will control floods any more than engineering solutions like dams and embankments have been able to.

Flood forecasting

Flood forecasting and warning plays a very significant role in saving human lives and livestock as well as in reducing the damage to portable property by informing the affected population in advance of the likely flood levels and their duration at specific places. It also helps in organizing timely rescue and flood-fighting measures in order to prevent/ minimize the damage to flood protection works like embankments. It is thus the recognized non-structural method of minimizing flood damage, although it cannot prevent or alter the occurrence of floods. In the GBM region, it is all the more significant in view of the facts mentioned earlier.

Effectiveness of flood forecasting increases with the increase in lead time, and forecasts should be given as much in advance as possible. This can be achieved by getting real-time information about rainfall as well as flows in upstream reaches of the rivers. Inter-country cooperation, therefore, is extremely useful in improving the efficacy and utility of flood forecasting. The present status in various countries is briefly summarized below.

In Nepal, at present during the monsoon, data from 12 rainfall and seven river-gauge stations are transmitted to Patna in India (Baweja and Rangachari 1997), although as far back as in 1987–88, a joint Indo-Nepal project for flood forecasting was envisaged, comprising the setting up of 45 rain-gauge stations and 20 river-gauge stations. All of these were to be equipped with automatic water-level recorders and self-recording rain-gauges. The observed data were to be transmitted to India on a real-time basis. However, this project has yet to become fully operational. It is reported (Yogacharya 1997) that the infrastructure at 43 stations has been completed and the balance were likely to be completed by the middle of 1998.

Bhutan is collecting and transmitting rainfall and river-flow data from selected sites on some of the tributaries of the Brahmaputra originating in Bhutan, like the Puthimari, Pagladiya, Manas, and Sankosh. At present, data from five stations are transmitted to Cooch Bihar in West Bengal and further data are transmitted to Barpeta/Nalbari in Assam from seven civil wireless and hydromet stations in Bhutan (Gyalshen 1997). The Bhutan Department of Power has also set up 70 Class C meteorological stations, four snow-gauging stations and 16 hydrological stations, mainly for planning purposes. There is considerable scope for improvement of observations, analysis, and transmission. Upgrading and strengthening of the present observation network in a phased manner is planned.

In India, flood forecasting has been done at 109 stations in the GBM region for a considerable time (Baweja and Rangachari 1997). Although upgrading and modernization of the system in a phased manner is under way, the track record so far of the timeliness and accuracy of forecasts has been very good. Under a bilateral arrangement, India is also transmitting actual and forecast river-level data to Bangladesh from five stations: Farakka on the Ganges, Goalpara and Dhubri on the Brahmaputra, Domohani on the Teesta, and Silchar on the Barak. In addition, rainfall data from Goalpara, Dhubri, Tura, Cooch-Behar, Siliguri, Jalpaiguri, and Agartala are also transmitted from India to Bangladesh.

Both India and Bangladesh at present are using similar technology for processing the data, statistical correlations, formulation of forecasts, and transmission. Both the countries have also sought assistance from DHI Denmark for setting up forecasting models like MIKE II and making them operational in a few forecasting systems.

In view of the large extent of the flood problem, for which there is no visible long-term solution, flood-forecasting networks in all the countries needs to be extended and strengthened considerably, including increasing the density and coverage of observation stations. While the upgrading and modernization will be a continuous long-term process to increase the accuracy and advance time (lead time) of forecasts, Indian experience indicates that it is always useful first to set up an observation and communication network to start forecasts with statistical co-relations, and later to upgrade the observation network and use forecasting models to incorporate other parameters.

Flood forecasting is again a non-controversial area of mutual cooperation not needing a heavy financial investment, but which will benefit the large population in this region by reducing their recurrent loss and misery. However, the progress so far is limited, and flood forecasting should now be taken up as a priority programme in regional cooperation. An international seminar held in Dhaka on 5–6 December 1997 was a good step forward, and the need for cooperation was agreed by all the countries.

Technical assistance in formulation of projects

As is well known, with the exception of Bhakra dam, where foreign technical assistance was sought, over the last 50 years India has planned, designed, constructed, and operated a number of water resources projects – irrigation, hydropower, and multipurpose – without external assistance. These projects include dams of various heights, barrages, weirs, and canal systems, as well as flood control works. On account of the wide topographical and hydrological variations in India, Indian engineers have gained considerable expertise in successfully tackling complex problems in different rivers. Their expertise and capability in development and management of water resources has also been recognized by international financial agencies like the World Bank, the ADB, the UNDP, etc. The track record of the safety of structures in India has also been extremely good as compared to other countries around the world.

The neighbouring countries in the region can, therefore, easily take advantage of the technical expertise available in India. Bhutan has already taken advantage of this for hydropower projects; similarly, Bangladesh has accepted in principle an Indian offer of technical assistance in the design of the proposed Ganges barrage in Bangladesh.

There has, however, been so far some reservation in Nepal about utilization of Indian technical assistance. In the 1970s and early 1980s, although it was decided that the entire survey, investigation, planning, and design of the Pancheswar project would be carried out jointly by

Indian and Nepali engineers, with the cost being borne by India, the proposal did not materialize, as Nepal wanted to use international consultants. In fact, the Pancheswar project is a simple project with no major geotechnical or other issues compared to many more difficult projects tackled by India.

The cost of international consultants is high – varying between eight and 10 times the cost of Indian consultants. In addition, a full knowledge of Himalayan geology, which in itself is highly complicated, is not available to the international experts. Indian experience for the last 20 years indicates that at times international experts did not take into account fully the local conditions or the economics as relevant to developing countries with limitations of financial resources, and tried to transplant ideas and methodologies adopted in developed countries, which are not necessarily useful and cost-effective for Asian conditions. This was one of the reasons why there was a considerable difference of opinion between the Indian engineers and the international consultants appointed by the government of Nepal for the preparation of a feasibility report on the Karnali (Chisapani) project. This also resulted in considerable delay in finalization of that report.

There was a similar example in Bangladesh, where, after the unprecedented floods of September 1988, a flood action plan was formulated by 11 donor countries and four multilateral agencies coordinated by the World Bank. The plan basically proposed the construction of a number of embankments with drainage sluices, etc., with a total estimated cost of about US$10 billion. The cost of the first study phase itself was US$150 million (Rs. 500 crores), 60 per cent of which was for the fees of foreign consultants. The plan provoked considerable reaction and resentment, not only among the flood-affected people but also in the scientific community in Bangladesh (Bangladesh Centre for Advanced Studies 1994). After a review of existing flood protection works and innumerable interviews with the affected people and a number of experts/scientists in Bangladesh, the Bangladesh Centre for Advanced Studies book clearly brought out that even the existing embankments had not resulted in an increase in agricultural production in the protected areas. It also revealed that almost 80 per cent of the embankments had maintenance problems and almost 66 per cent were subject to frequent breaches. On the other hand, the embankments had resulted in a permanent loss of annual fish production of 45,000 tonnes by preventing migration of fish and natural restocking. Fish production is one of the main means of subsistence in Bangladesh. The book clearly highlights considerable opposition to a plan prepared by foreign experts from all concerned; according to Bangladesh, the plan will cause great potential damage to the Bangladeshi economy by importing the views of foreign experts without fully understanding the local problems and without the involvement of the people to be benefited.

In India also, in some of the projects assisted by the World Bank, there were similar problems, although there were no major changes in the planning and designs which were produced by the Indian engineers.

Taking all these factors into account, it will be in the interest of the region to avail itself of the technical expertise already developed in the region, which will also be economically cost-effective. At the most, since some programmes and projects will be financed by other countries/ international agencies which insist on the use of foreign consultants, a broad review of the project formulation can be carried out by international consultants and their suggestions considered in finalizing the project reports.

Since the sizing of a project may create differences of opinion, leading to delays in finalization, it will be desirable to work out different scenarios of development along with their economics, so that after discussions and mutual agreement the exact scope of the project and techno-economics can be expeditiously finalized without resorting to further investigations and studies.

Joint implementation and operation

As already indicated, the existing Chukha hydro-electric project in Bhutan was jointly implemented and is now being successfully operated. A similar arrangement of having a specific project authority for implementation and operation of each of the proposed major hydro projects in Bhutan, with equal representation of both India and Bhutan, has been agreed to. In the case of the proposed Pancheswar project, a similar joint Pancheswar authority is envisaged. Since projects in Nepal and Bhutan will have varied and appreciable repercussions in India, apart from the fact that India will be the major purchaser of the hydropower to be produced, it will be essential to have joint control for implementation and operation of these projects, although they will be located within the sovereign territories of Nepal and Bhutan. This aspect will assume particular significance in respect of projects in Nepal, as most of these will be multipurpose – having irrigation benefits in both India and Nepal, generation of hydropower to be consumed in India, and flood control benefits, again in both Nepal and India. It is also likely that apart from purchasing the power at mutually agreed tariffs, India may have to share a portion of the cost of projects which will have specific and identifiable irrigation and flood control benefits in India, depending on the quantum of benefits. In such cases, joint planning, implementation, and operation will create confidence in India that the planned benefits will be available on a sustainable basis in the future. The exact functions and type of composition of the joint machinery can, however, vary with each project; but agreement on this should not delay the formulation of detailed project reports. After the successful experience of Pancheswar when it is implemented, it will be easier to have such joint arrangements for other projects in Nepal.

Management and equitable sharing of lean flows

India-Bangladesh

A long-term (30 years) treaty on sharing of the lean flows of the Ganges between India and Bangladesh has recently been concluded. Similar arrangements need to be evolved for the Teesta and some of the common small rivers flowing from India into Bangladesh (tentatively the number is 54). It has been recognized that the available flows in the Teesta during the lean season are not sufficient to meet the reasonable needs of both India and Bangladesh. A number of studies carried out earlier in India indicated that it is not feasible to construct a major dam on the Teesta river in India, because it would have a very limited life thanks to the tremendous silt load carried by the Teesta all year round. Therefore, the development in the Teesta region in both India and Bangladesh has to be mainly "run of the river". Already two barrages, one in India and the other in Bangladesh, have been constructed for diverting the available flow to provide irrigation facilities in both countries. A treaty similar to the Ganges treaty on sharing the lean flows, after taking into account the realistic needs of both countries, will have to be evolved; the governing principle will be that during periods of flow inadequate to the requirements of both, the shortages will have to be equitably shared by both the countries.

On the smaller common rivers, again, the possibility of constructing storage dams in India will be very limited, and full utilization of available surface waters may not be possible. Here again, sharing of flows with run-of-the-river development will have to be resorted to. So far, no significant studies have been carried out in either India or Bangladesh for determining the actual requirements of optimal use of the waters of these rivers. Therefore, it will be desirable that instead of tackling the problem of all the common rivers, initially detailed studies should be made in respect of a few identified rivers; meanwhile the broad studies of requirements of other rivers can continue.

India-Nepal

As regards sharing of flows between Nepal and India, since in the past there was no significant irrigation development in Nepal, the flows in the lean season coming to India from various rivers in Nepal have been utilized for irrigation in India for a number of years by constructing barrages and canal systems – particularly on various tributaries of the Ganges like the Kamla, Bagmati, Gandak, etc. Nepal has now been developing projects on some of the smaller tributaries – the Rapti, Ghagra, Kamla, etc. – for providing irrigation in Nepal. As strictly such a use by Nepal cannot be totally opposed on the basis of the recognized principle of disturbance in the established riparian rights, it should be possible to evolve some arrangements where equitable sharing of lean flows can be agreed. Besides, in most of the tributaries there is scope for constructing medium to large storage dams in Nepal which will store the flood waters and thus make more waters available for utilization in both Nepal and India. Such projects, which may not have a very long gestation period as well as not requiring substantial investment, can be implemented early; again, sharing of costs by both countries will not only sustain the existing benefits being derived by India but also enhance these. Since the overall requirements of consumption in Nepal will be comparatively much smaller, as a principle, in any such storage projects the full requirements of Nepal should be met. Thereafter, the unutilized portion of water from Nepal, which naturally has to flow through India, can be developed and utilized by India in the best manner possible. In some cases, apart from irrigation it may also be possible to develop navigation, through either the rivers or the canal systems; this again can be studied while detailed planning of the projects is carried out.

Once the genuine desire for and determination of mutual cooperation is established, it should be possible to evolve satisfactory solutions on the basis of a flexible "give-and-take" attitude. Continuous dialogue and interaction at both technical and political levels will, however, be necessary.

Augmentation of flows

Demand scenario

The irrigation potential of the Ganges basin in India is estimated to be of the order of 61 million ha. The potential already created is about 20 million ha. As pointed out earlier, the present irrigation is essentially based on river diversions (run-of-the-river development), as there are no storage dams possible on the tributaries of the Ganges in India. The low flows at present available in the various tributaries and the main river are totally inadequate to meet the requirements of the balance of irrigation potential of about 41 million ha. On the other hand, with progressive utilization of lean flows in Nepal, there will be some reduction in the existing low flows in the Indian territory. The present population in the Ganges basin portion of India is about 400 million, and this is expected to increase to about 550 million by the year 2010. Consequently, the demand of water for various purposes, including domestic, municipal, livestock, and agriculture, will be progressively increasing every year.

In Nepal, it is estimated that the present irrigation is of the order of about 1 million ha and the total population is about 20 million. According to a master plan prepared in Nepal, it may be possible to provide irrigation to an area of about 2.8 million ha in Nepal and in addition about 4.2 million ha in India by constructing 13 dams with a total cumulative storage of around 7.7 million ha-m.

In Bangladesh, the potential area requiring irrigation depending on the waters of the Ganges is estimated to be about 1.31 million ha. The area actually irrigated at present is not known.

In the Brahmaputra and Meghna river basins in India, the in-basin utilization is estimated to be only about 4 per cent of the available waters. Similarly, the requirement for consumption in Bangladesh is also comparatively very limited.

Thus while there is a shortage of flows in the Ganges basin to meet the future requirements of Nepal, India, and Bangladesh, there is surplus water available in the Brahmaputra river.

Ways of augmentation

The term "augmentation" connotes augmentation of flows during the lean season. It does not imply overall augmentation of water resources, as these are fixed. There are three different possible means of augmentation:

- creation of major storage facilities by dams and releasing the stored waters during the lean season;
- utilizing a portion of the stored waters behind the dams for trans-basin diversion to adjoining river basins;
- constructing diversion structures across the river (without any storage) for transfer of the surplus flow of one river system to another by canals and river structures like barrages en route.

Within the Ganges basin

As already discussed, there are no possibilities of constructing dams on the main Ganges river or its tributaries in India – particularly northern tributaries. Storage facilities on the southern tributaries have already been built, and these waters are being almost fully utilized within the Ganges basin in India. However, there are great possibilities for constructing major storage facilities on the northern tributaries of the Ganges in Nepal. According to the master plan prepared by Nepal, 13 dam sites on various tributaries are possible, with a total storage capacity of about 7.7 million ha-m. Of these, the major storage dams are Pancheswar, Karnali, and on the Kosi, which together will provide a live storage capacity of about 5.5 million ha-m. With the construction of these facilities, it will be possible to augment the lean-season flows in various tributaries to provide irrigation benefits both in Nepal and in India. However, excepting the proposed high dam on the main Kosi, construction of other dams is not likely appreciably to augment the lean flows in the main Ganges at Farakka, as most of the stored waters will be utilized in Nepal and India. The Kosi is one of the largest tributaries of the Ganges and the consumption requirements in the Kosi sub-basin in both Nepal and India are limited. As such, the stored waters in the Kosi dam could be utilized for augmentation of lean flows in the Ganges upstream of Farakka, simultaneously using the conveyance of waters for navigation between Nepal and India. Thus the only reliable source of augmentation of the Ganges flows will be the Kosi dam if and when implemented, although Bangladesh has been pleading for construction of dams on various tributaries of the Ganges in Nepal. Such dams will no doubt be beneficial to both Nepal and India for multipurpose use, but are not likely to augment the main Ganges flows at Farakka in any sizeable way.

Transfer from the Brahmaputra

There are surplus waters in the Brahmaputra, and the flows even during the lean season at Jogigopa in India vary from 126,000 cusecs in January to 200,000 cusecs in April. Further, there is a lag in this lean-season flow compared to the lean-season flow in the Ganges, which occurs during the months of March and April. Because of these facts, various possibilities of diverting the flows of the Brahmaputra into the Ganges upstream of Farakka have been under consideration.

One of the earliest proposals suggested by India was to construct a barrage on the Brahmaputra at Jogigopa in Assam (India), diverting about 100,000 cusecs of water through a 324-km-long link canal crossing Bangladesh territory and then reentering India and outfalling upstream of Farakka. However, such a proposal was not found acceptable by Bangladesh, for technical and other reasons.

Another proposal was broadly studied, wherein the link canal instead of passing through Bangladesh would traverse entirely within Indian territory from the Jogigopa barrage and connect the Ganges through the Teesta and Mahananda rivers in India. This would also provide large irrigation benefits in Assam and West Bengal. However, such a proposal involves considerable pumping, as the waters have to be lifted in different stages from Jogigopa up to the Teesta river, whence it flows by gravity through the Mahananda to the Ganges. It was considered that consumption of such a large quantum of power, apart from the dim prospects of its availability, along with very heavy annual operation costs may not render the proposal techno-economically viable.

The third alternative (Char 1997b) is a diversion channel from the proposed dam to be constructed on the Sankosh river in Bhutan, where apart from generating 4,000 MW of power, the channel will enable a discharge of about 12,000 cusecs to be diverted through a link canal falling into the Teesta barrage in India and later augmenting the flows of the Ganges at Farakka. This proposal is yet to be studied and fully investigated in India by the Central Water Commission and the Ministry of Environment and Forests.

While trans-basin diversions of water have been attempted in other countries, and even a proposal for linking some of the major rivers in peninsular and southern India are under study, there is considerable opposition to such proposals from the environmentalists, apart from the heavy capital investment. It is thus not clear whether such mass water transfer proposals, although technically feasible, will be implemented at all.

Interconnection within Bangladesh

There is a possible alternative of diverting the Brahmaputra through a link canal within Bangladesh itself. Such a proposal would be the shortest link and would give substantial irrigation benefits even in the Teesta region in Bangladesh. In the past Bangladesh did not favour such a proposal on the grounds that there is no surplus water in the Brahmaputra. However, it is now understood that a Brahmaputra barrage within Bangladesh territory is likely to be investigated and studied by Bangladesh. This barrage, coupled with the proposed Ganges barrage now being planned in Bangladesh, could substantially improve the surface irrigation facilities in that country.

In view of these various facts, while the studies on possible alternatives can continue, the viable solution in the near future appears to be the construction of barrages on both the Ganges and the Brahmaputra in Bangladesh itself. In this connection it also needs to be noted that the investment required in the construction of major dams in Nepal will be very high and will be beyond the financial capabilities of both India and Nepal. Even with some international financial assistance, realistically, these appear to be projects in a very distant future. It is therefore necessary to take a pragmatic view of the most feasible proposal of augmentation, and that which is likely to give benefits in the near future.

Conclusions and recommendations

Project reports

After necessary investigations and studies, the project reports of major storage dams in Nepal should be formulated for different scenarios of development and finalized in consultation with India. A start has already been made with the Mahakali treaty. For implementation of this treaty, the Pancheswar project needs to be finalized expeditiously. This should be followed by the Karnali and Kosi projects, apart from other smaller projects.

The project reports on the Ganges and the Brahmaputra barrages in Bangladesh need to be finalized early with the technical assistance already offered by India.

Flood forecasting

It is absolutely necessary to extend, strengthen, and modernize the flood forecasting system and the network covering Nepal, Bhutan, India, and Bangladesh for the benefit of all countries. This proposal is non-controversial and will need comparatively little investment – it should be given top priority.

Interconnection of power systems

Studies of interconnection of power systems in India and Nepal and their techno-economics need to be carried out to enable implementation of the necessary transmission network for transfer of power first from Pancheswar and then from other future projects in Nepal.

Barrages in Bangladesh

In view of the fact that no other augmentation proposal is likely to materialize in the near future, it would be in the interest of Bangladesh to undertake the construction of barrages on both the Ganges and the Brahmaputra in their territory as early as possible.

Sharing of lean flows

Earnest efforts for an amicable settlement between India and Bangladesh should continue for sharing of the lean flows of the Teesta and other small common rivers.

REFERENCES

Agarwal, A. 1991. *State of India's Environment: A Citizens' Report.* New Delhi: Centre for Science and Environment.

Bangladesh Centre for Advanced Studies. 1994. *Rivers of Life.*

Baweja, M. L. and R. Rangachari. 1997. *Indo-Bangladesh-Nepal Cooperation in Flood Disaster Management.* New Delhi, Centre for Policy Research.

Char, N. V. V. 1997a. *Integrated Water Resources Development of Ganga, Brahmaputra and Meghna River Systems – International Dimensions.* New Delhi: Ministry of Water Resources.

Char, N. V. V. 1997b. *Augmentation of Water Resources in the Ganga Basin.* New Delhi: Ministry of Water Resources.

Gyalshen, Wangdi. 1997. *Country Report on Flood Forecasting and Warning Status in Bhutan.* Thimphu: Ministry of Planning.

Leopold, L. and T. Maddock. 1961. *Flood Control Controversy.* Washington, DC: US Geological Society.

Prasad, T. 1997. *Integrated Water-Based Development of the GBM Region – Indo-Nepal Aspects.* New Delhi: Centre for Policy Research.

Rangachari, R. 1997. *Water-based Integrated Development of the GBM Region.* New Delhi: Centre for Policy Research.

Sambamurty, M. K. 1997. *The Eastern Power/Energy Grid.* New Delhi: Centre for Policy Research.

Shah, R. B. 1996. "Water Resources – Myths and Realities", Dr Kanwar Sain memorial lecture, New Delhi.

Yogacharya, Kiran Shankar. 1997. *Status Paper on Flood Forecasting and Warning System in Nepal.* Kathmandu: Ministry of Irrigation.

3

Constraints and opportunities for cooperation towards development of water resources in the Ganges basin

A. T. M. Shamsul Huda

Introduction

River systems constitute distinctive hydrological units and nature wants them to be treated as such. Unfortunately, there are very few major river systems that flow through one single country; most are transboundary rivers. In such a situation, basinwide planning for water resources' development and management would be the most logical approach by the co-riparian countries. However, the political, social, and cultural dispositions of these countries, supported by the current political doctrines, have not been very conducive to the expected level of cooperation.

The Ganges basin: The physical conditions in Bangladesh

The lack of cooperation has been particularly acute in the case of South Asia, and nowhere has this been more acute than in the Ganges basin. The mighty river Ganges is about 2,550 km long. After taking off from the Gangotri glacier near the Indo-Chinese border, it passes through Nepal, India, and Bangladesh before merging with the Bay of Bengal. The basin accounts for 104.74 million ha, of which 86.14 million ha fall within India, 14 million ha in Nepal and 4.60 million ha in Bangladesh. Like any major river system, the significance of the Ganges lies in its very extensive network. The Ganges receives large volumes of water from Nepal through its three left-bank tributaries – the Karnali, the Gandaki, and the Kosi. It distributes water to large parts of Uttar Pradesh, Bihar, Assam, and West Bengal, and through a network of distributaries to north-west and south-west parts of Bangladesh.

The tributaries of the Ganges in Nepal are snow and glacier fed. A large part of their drainage area is covered by snow and glaciers throughout the year. The major floods experienced in the Himalayas are mainly due to glacier lake outbursts or cloud outbursts. Rockslides and landslides, which are common in Nepal, also aggravate flooding problems by reducing river capacity or even by temporarily damming the rivers. Moreover, there have been a number of instances of extraordinary floods due to outbursts of temporary dams erected by rockslides and landslides in the rivers (Bangladesh-Nepal Joint Study Team 1989).

The Ganges basin is experiencing intensive erosion due to natural geomorphologic conditions, intense rainfall, frequent seismic activity, and inappropriate land-use practices. The Ganges river system carries a large quantity of sediment and delivers this each year into the Bay of Bengal.

Geography has destined Bangladesh to face the unenviable situation of too much water during the monsoon and too little during the dry season. The fact that Bangladesh occupies only 7 per cent of the Ganges basin shows the country's lack of control over the inflow of water that determines the geophysical conditions. The various manifestations of these conditions are enumerated in the following sections

Flood and drought

The country's unique physical setting makes it extremely vulnerable to both flood and drought. The annual minimum flow of rivers near the Bay of Bengal is only 5 per cent of their peak monsoon discharge. The high-water levels in the major rivers during monsoon cause drainage congestion and over-bank spillage of their tributaries, which is often exacerbated by runoff from coincident high local rainfall.

Bangladesh generally enjoys a subtropical monsoon climate. It receives a normal annual rainfall ranging from 1,200 mm in the extreme west to as high as 5,800 mm. About 81 per cent of Bangladeshi rainfall occurs in the summer humid period – May to October.

The country, being the lowest riparian, offers an outlet for the entire volume of flow of all upstream major rivers. The magnitude of flood damage is of catastrophic proportions when the flood waves from the Ganges, Brahmaputra, and Meghna combine with rainfall all over the country. Rainfall of long duration in the monsoon season often generates water volume in excess of the local drainage capacity, causing localized floods.

Certain parts of the basin are also subjected to drought during the monsoon season, which adversely affects the cultivation of summer *kharif* crops. The upstream withdrawal of water during the dry season has further aggravated the situation, giving rise to the need for supplementary irrigation.

Siltation

Each year, some 2.5 million imperial tons of sediment are brought down by the Ganges and transported through the area to the Bay of Bengal (Bangladesh Ministry of Water Resources 1998). A part of this sediment load is deposited on the floodplains during the flood season, gradually changing their topography and drainage conditions. Continued siltation has reduced the efficiency of Mongla port. Low flows in the Ganges have seriously affected navigation on the rivers in the Ganges-dependent areas, with 160 km of the Ganges, 70 km of the Gorai, and the Modhumati and Passur rivers suffering from reduced flows and siltation.

River-bank erosion

The floodplains and coastal delta are in a constant state of slow morphological change. The large seasonal variation in river flow results in a varying sediment transportation capacity and causes river-bank erosion, migration of river banks, and meandering river channels. Recent satellite-image studies of the Ganges-Brahmaputra-middle Meghna rivers under the Flood Action Plan show that 106,300 ha were lost to erosion, while only 19,300 ha accreted, over the period 1982–92. The net area of 87,000 ha lost to erosion is equivalent to an annual erosion rate of 8,700 ha, most of it agricultural land. Erosion of border river banks has the serious implication of losing lands to a neighbouring country. River-bank erosion has significant economic and social impact. The loss of land, crops, and property has led to landlessness and impoverishment for thousands of households (World Bank 1996).

Salinity intrusion

After commissioning of the Farakka barrage in West Bengal, the net availability of surface water to Bangladesh via the Ganges during the dry season has been reduced beyond critical levels. The impact of this reduction is demonstrated by a marked increase in salinity in south-western Bangladesh. The southern part of this affected region is subject to tidal action from the Bay of Bengal, and saline intrusion from the tides is normally pushed back by upland flows of fresh water from the Ganges through its tributaries, among which the Gorai plays a vital role. Due to flow reduction since 1975, tidal limits and the salinity front have moved northward. Around Khulna, some 146 km upstream from the Bay of Bengal, the salinity (in terms of electric conductivity) has increased from 380 micro-mhos/cm in the pre-barrage diversion period to 29,000 micro-mhos/cm. This is a major problem in over 25,900 km^2 of Ganges-dependent area (GDA) and is causing both short- and long-term problems in crop production, fisheries, forestry, power generation, industrial development, health care, and domestic water supply (Bangladesh Ministry of Water Resources 1997).

Arsenic contamination

Recently, arsenic contamination in groundwater, far in excess of the World Health Organization's drinking water provisional guideline value of 0.01 mg/l, has been detected in several districts of Bangladesh. Of these, Rajshahi, Jessore, Meherpur, Kushtia, Khulna, Bagerhat, and Faridpur have been identified as the highly affected districts. This development has posed a serious threat to the health of people already affected and the people at risk.

The cost and consequences of lack of cooperation

The Ganges basin is one of the most densely populated and poverty-stricken areas in South Asia. It is also one of the most richly endowed in the world, but its potential is the least exploited. The cost and consequences of lack of cooperation so far as these relate to Bangladesh have not been systematically explored. However, by looking at a few key sectors like agriculture, forestry, and health we can have some idea of the extent of damage being caused to the economy of Bangladesh.

Agriculture

The south-west region situated in the Ganges basin is deficit in food production. Of the total area of 40,500 km^2, only 62 per cent is cultivable and 10 per cent is covered by mangrove forest, the Sundarbans, an area of immense environmental importance. The area is predominantly low lying with a maximum elevation in the north of about 14 metres above sea level. The eastern half tends to be flood prone, while the west suffers an acute shortfall of water in the dry season. Groundwater is available in limited quantity in the northern half, but there is hardly any of usable quality in the south. Saline water intrudes from the southern coast as deep as 185 km upland, and the salinity itself has increased over the years as inflows from the Ganges have reduced.

The Ganges-Kobadak project, a major gravity system irrigation project, was built to irrigate 125,000 ha, but actual coverage for Kharif II could not be raised above 80,000 ha.

Bangladesh has set a target for production of 25 million imperial tons of food grain per annum by the year 2010 to keep pace with the growth of population. To contribute to this national effort, the GDA has to raise production through expanded irrigation. In the absence of a sizeable surface irrigation programme, groundwater irrigation has grown very rapidly in the GDA. However, this growth may not be sustainable in the long run. The following developments have raised concern about the unfettered growth of groundwater irrigation:

- expansion of groundwater use entails the lowering of groundwater tables, which affects domestic supply wells, the sustainability of ponds, and, in the long term, equity and social equilibrium amongst user groups;
- too much extraction of groundwater may have something to do with the release of arsenic into water supplies;
- studies have shown that at the current rate of expansion, groundwater exploitation would reach its upper limits in another five to 10 years' time (Master Plan Organization 1991).

The fundamental issue for agricultural development in the GDA that emerges from the above is the need for ensuring fresh water availability and balanced use of groundwater and surface water. The key to doing all that is the management of the Ganges water in a manner that will supply fresh water through its distributaries. Preliminary studies have indicated that this can be done by diversion of Ganges water during the dry season by constructing a barrage on the Ganges. Undertaking such a large project needs the active support and cooperation of India as an upper riparian. This has not been possible in the absence of a long-term water-sharing arrangement.

The delay in arriving at some long-term understanding on water sharing has very adversely impacted the agriculture of the GDA in a number of ways. Soil moisture depletion, lowering of the groundwater table, increase in salinity content in water, and, above all, reduction of acreage for cultivation due to the shortfall in surface water irrigation have all contributed to stagnation in agricultural growth. One study estimated the direct loss to agriculture due to these reasons at about Taka 2,000 million per annum (US$1 = Taka 44) at 1991 prices (Nishat 1996).

Forestry

The GDA is rich in forest resources and is home to the Sundarbans, the largest mangrove forest in the world. The forest is spread over an area of about 58,000 ha, interlaced with 170,000 ha of tidal channels and rivers. With a vast range of plants and animals, including the Royal Bengal tiger, it is a rich source of biodiversity. In recognition to its value, UNESCO in 1997 designated 140,000 ha of the forest as a World Heritage site.

In recent years, the phenomenon of "top dying", which afflicts the sundari trees and affected 17 per cent of the forest in 1985, is thought to be caused by a synergistic reaction to flows characterized by increased salinity and higher levels of industrial pollution. Damage to the sundari forests threatens the integrity of this important habitat and the species dependent on it as well as its economic potential.

The Sundarbans act as a buffer against tidal waves. Substituting the forest with artificial defences may cost around Taka 16,000 million with Taka 320 million as annual maintenance costs (Bangladesh Ministry of Water Resources 1998). Moreover, about 90,000 people depend on the area for their livelihood. The Sundarbans can only be saved by reversing the process of degradation caused by excessive salinity and other human-induced changes. If fresh water flow and the flood regime are restored, the mangrove forests and the ecosystem can be revitalized.

Health

Over the last three decades, environmental decadence in the GDA has adversely affected the health of the people of the area. The fall in groundwater levels and reduction in dry-season flow of the Ganges have adversely affected the quality of water. Scarcity of pure drinking water has caused a rise in the incidence of various enteric diseases in the region.

Acute arsenicosis has been detected in the far west of the south-west and the Paksi region in particular, and research is now being conducted into the viability of alternative drinking water sources, both from groundwater and surface water sources.

As has already been pointed out, there is no systematic study of the losses sustained by Bangladesh due to failure to take action to reverse the environmental degradation set in motion in the GDA. However, by even a rough estimate, the damages done to the economy of Bangladesh by not taking any action to mitigate the physical conditions in the south-west are enormous. One recent attempt to calculate that put it at about Taka 108,500 million (Nishat 1996). The figure must be taken to be tentative, but it gives an idea of the cost of not taking appropriate decisions at the right time.

Opportunities for cooperation in the Ganges basin

In absolute terms, the total quantum of water available in the catchment area of the Ganges-Brahmaputra-Meghna basins is enough to meet the requirements of all the co-basin countries. Though there is wide seasonal and spatial variation in availability of water, this can be overcome through proper planning and management of the waters in the region.

Lacking this much-needed cooperation and out of desperation, Bangladesh has ventured in the past to manage waters within its own boundaries. One such major effort was the launching of the Flood Action Plan in 1990, aimed at controlling the adverse effects of devastating floods in Bangladesh. The country was driven to such a desperate action following the floods of 1988. On completion of a number of studies by 1996, the idea of total control of floods was abandoned, due largely to the fact that for any effective flood control measure cooperation from the upper riparians was considered essential (Bangladesh Ministry of Water Resources 1997).

The Ganges water-sharing treaty of 1996 signed between Bangladesh and India is indeed a landmark treaty, and has resolved a long-standing irritant between the two countries. However, the fact remains that the dry-season flow of the Ganges at Farakka is insufficient to meet the demands of Bangladesh and India. The demand for water from the tributaries of the Ganges in Nepal is also increasing. There is thus an urgent need to augment the flows of the Ganges.

The characteristics of water resources in the Ganges basin in terms of spatial and temporal distribution are such that these can be used to the mutual benefit of the co-basin countries. Flooding and reduction of flow during the dry season that bedevil south-western Bangladesh can be taken care of by Nepal by regulation of flows upstream.

The concept of storage reservoirs in the upper reaches of the Ganges has engaged the attention of concerned experts for a long time. Preliminary studies conducted in 1978 by Bangladesh and updated in 1983 revealed the possibility of constructing seven storage reservoirs in Nepal along with a number of storage dams in India. Studies conducted by Nepal at different levels (master plan, pre-feasibility, and feasibility) have identified about 30 reservoir sites. The total storage capacity of high-dam projects identified so far in Nepal is of the order of 82 billion cubic metres of live storage that would regulate over 95 per cent of the total annual flow (Bangladesh-Nepal Joint Study Team 1989).

The storage reservoirs can hold the vast monsoon runoff within Nepal and, since all this volume of water has to pass along the Ganges into the Bay of Bengal, they will play a very significant role in mitigating the adverse affects of severe floods in India and Bangladesh. Besides flood mitigation, the stored water would go a long way towards augmenting the dry-season flow of the Ganges. Augmentation potential in Nepal during the dry season can range from 2,400 to 4,950 cumecs. These incremental flows alone are over four times the present lean-season flows in the Ganges at Farakka. A single storage facility such as the Karnali project alone has the augmentation potential to more than double the existing low flow of the Ganges.

Flow regulation has the further potential of expanding dry-season irrigation and the possibility of river navigation. When fully developed, regulated flow can be used to irrigate 27 million hectares of land. Since the availability of agricultural land is very limited in Nepal, most of this surplus water can be used by water-hungry co-basin countries like India and Bangladesh.

Water control in the upper reaches also opens up the opportunity for inland river navigation throughout the entire channel. Nepal is a land-locked country, and the north-eastern states of India also lack direct access to other parts of India. All of them need transit facilities through Bangladesh for access to the sea, or to other parts of their own country. Development of reservoirs help these possibilities in at least two ways. First, flow augmentation can ensure round-the-year navigation by improving the barrage for inland water transportation. Secondly, the Ganges could provide a riverine transit from Nepal to a seaport. Similar transit opportunities may be available to the north-eastern states of India.

The Mahakali treaty and the Ganges water-sharing treaty signed by India with Nepal and Bangladesh respectively are bilateral in nature. These have been signed after many years of patient negotiations. At this stage, the concept of basinwide planning may not be immediately attainable, though that should be the goal to be achieved by all the co-riparian countries. The transition from enlightened bilateralism to commercially oriented multilateralism may not be too far away: but the distance needs to be traversed carefully by overcoming the obstacles on the way one by one.

Approaches to cooperation

The foregoing discussion has tried to bring into focus the imperatives of cooperation for water resources development among co-riparian countries. The cost of not doing that has also been highlighted by citing the current status of a few important sectors in the GDA in Bangladesh.

The history of interaction between co-riparian countries does not contain many success stories, but efforts at such cooperation must continue. In South Asia, the full implementation of the Indus basin treaty of 1980 between India and Pakistan has been beneficial to both and has removed for good one major source of tension between them.

Mutual distrust, lack of vision, and pure narrow-mindedness appear to be the major constraints for an integrated plan for water resources development. These constraints cannot be removed overnight, nor can they be ignored. In the light of recent experience, there can be a variety of approaches to pursue the long-term goal of substantive cooperation between and among co-riparian countries.

The two bilateral treaties related to the Ganges basin are landmark treaties in the sense of cooperation among two co-riparian countries. A couple of years ago, this framework did not exist. The relationship created by the treaties is new and is still shrouded by disbelief and mistrust. In both the cases, the political will and statesmanlike vision of the respective leaders clinched the deals. Civil servants and the water experts have to come to terms with these sudden realities by shaking off their long-held rigid technical orthodoxies. In the immediate future, no major achievements in basinwide planning and development of water resources are visualized. However, there are a number of approaches to bilateral cooperation through which, over time, we could achieve a much greater level of cooperation among all co-basin countries.

Confidence-building

The benefits of cooperation are easy to appreciate if they are discussed and projected. The relationship among the co-basin countries has been influenced by history, and their unequal size, resource base, and capacity. India, as the big neighbour, would have to take steps to build confidence in her smaller neighbours. This can be done by scrupulously following the provisions of any treaty. In the case of Bangladesh, for example, the delivery of water during the dry season as per schedule would be a positive measure toward such confidence-building.

Picking the least difficult

In the expansion of bilateral cooperation, there will be a host of issues needing attention. Some of these will be very complex, some a little less so, while a few may be not that difficult. The ones in the last category are those most likely to succeed. It is a good idea to target these activities and set examples of good neighbourly cooperation. After the Ganges treaty, the Joint Rivers Commission charged with the implementation of the treaty decided in its first meeting, held after a lapse of seven years, to start work towards linking the two embankments on the right bank of the Teesta falling within the respective jurisdictions of the two countries. This had been a pending issue for a long time, although its implementation was not likely to harm any side. Rather, when completed this season, it will save the thousands of people living on both sides of the border from the havoc of flash floods.

Supporting activities

Misinformation breeds mistrust. There is a tremendous lack of information as to the potential benefits of cooperation among co-riparians on a commercial basis. Transferring benefits on commercial terms, particularly from a small country like Bangladesh to a big neighbour like India, is viewed by some members of the public with suspicion. There should be regular exchanges of visits by scholars, academics, and officials of the concerned countries and open discussion on the imperatives of cooperation. The recent Track II activities by independent scholars, water experts, and private research foundations have contributed significantly in creating awareness about the problems and prospects of water resources development in the Ganges basin. On their way to the Ganges water-sharing treaty, both the governments concerned received much-needed intellectual and moral support from these organizations. With the signing of the two treaties, their activities should neither diminish nor cease; rather, they should be geared up to maintain the momentum and help implementation. Successful implementation of bilateral treaties shall pave the way for expansion of cooperation with other countries.

Transparency and sharing of information

Even when programmes are discussed and finalized on a bilateral basis, it is important to keep other co-riparians informed of every important step taken. It is always better that the concerned government gets the information from its counterpart rather than getting the news of developments in a common basin from newspaper columns.

An agenda for action

An important area where collaboration between neighboring countries could yield substantial benefit is integrated basin and inter-basin water management. This is an overriding consideration for Bangladesh because, for a greater part, its river basins lie in India and Nepal. Basin management is a complex task and requires extensive study of the hydrological and morphological characteristics of the terrain that underlies the territory of all three countries. The following action plan may show the steps towards the evolution of an effective water resource management system for the entire Ganges plain.

- Each of the three countries, Bangladesh, India, and Nepal, must establish a system for exchange of information and data on relevant aspects of hydrology, morphology, ecology, changing watershed characteristics, cyclone, drought, and flood warning, etc., to help each other understand current and future problems for short-, intermediate, and long-run planning.
- Each of the three countries should designate a technical agency; these could consult each other on matters of water management stretching across national boundaries.
- A set number of planning meetings should be held between the technical agencies each year to discuss matters of mutual interest on water management.
- A joint assessment of international river resources should be made by Bangladesh, India, and Nepal for understanding the overall basins' potential.
- The three countries should work towards establishing joint river basin organizations to act as central clearing houses for planning, water-sharing, and the preparation and execution of jointly owned projects.
- Concerted efforts should be made by all three countries to manage the catchment areas, with the help of afforestation and erosion control for watershed preservation and reduction of land degradation.
- Concerted efforts should be made by the three countries for preventing chemical and biological pollution of the rivers flowing through these countries, by managing the discharge of industrial, agricultural, and other pollutants generated by human action.

REFERENCES

Bangladesh Ministry of Water Resources. 1997. *Bangladesh Water and Flood Management Strategy: An Update Following the Signing of the Ganges Water-Sharing Treaty*. Dhaka: Ministry of Water Resources.

Bangladesh Ministry of Water Resources. 1998. *Socio-Economic and Environmental Status of the Ganges-Dependent Area with Reference to Water Resources Development*. Dhaka: Ministry of Water Resources.

Bangladesh-Nepal Joint Study Team. 1989. *Report on Mitigation Measures and*

Multipurpose Use of Water Resources. Dhaka and Kathmandu: Ministry of Irrigation, Water Development, and Flood Control, Dhaka and Ministry of Water Resources, Kathmandu.

Master Plan Organization. 1991. *National Water Plan*. Dhaka: Ministry of Irrigation, Water Development, and Flood Control.

Nishat, Ainun. 1996. "Impact of Ganges Water Dispute on Bangladesh", in Asit K. Biswas and Tsuyoshi Hashimoto, eds, *Asian International Waters: From Ganges-Brahmaputra to Mekong*. Bombay: Oxford University Press.

World Bank. 1996. *Bangladesh River Bank Protection Project Staff Appraisal Report*. Washington, DC: World Bank.

4

Brahmaputra: Issues in development

A. D. Mohile

Introduction

Background and scope

The South Asian region is one of the most populated regions of the world. High population density coupled with low per capita income and a predominantly rural agricultural-based population are distinguishing features of this area.

The general geographic region which can be called the Bay of Bengal basin, comprising the various river systems flowing into the Bay of Bengal, is a typical part of the South Asian region. This area is marked by some of the most fertile alluvial plains, and also high mountains and deltas. The region has considerable water resources which are rather unevenly distributed. Given the agrarian nature of the economy, water development is basic of the development of the area.

The Ganges-Brahmaputra-Meghna system is the main basin, covering approximately 1.65 million square kilometres. It has an average water potential exceeding 1,000 km^3 per annum evenly distributed in a region which has the world's highest annual rainfall as well as semi-arid areas. The Ganges, the Brahmaputra, and the Barak-Meghna are separate subsystems in the combined river system, but they have a common delta system providing a common terminus into the sea and have considerable potential for spatial integration, both through possible physical links be-

Table 4.1 Countrywise length and catchment area of the Brahmaputra

Country	Length (km)	Catchment area (km^2)
China	1,625	293,000
India	918	195,000
Bangladesh	363 (up to Bay of Bengal)	47,000 (up to Goalando)

Table 4.2 Average annual flow of the Brahmaputra

River site	Period	Average annual yield (billion m^3)
Bechamara	1976–1983	269
Pandu	1955–1982	494
Jogighopa	1955–1957	537

tween the subsystems and in the possibility of shifting the water demand of the common delta to the water-rich subsystem in order to support larger development in other areas.

Such an approach towards the integrated development of the basin and the general region through the development of the Brahmaputra is the theme of this chapter. The local demands of the areas of the Brahmaputra subsystem need to be met as a priority before considering integrated basin or regional developments. For this purpose, a projection of the demands of the north-east region of India is presented here.

The author is of the view that even while the basin serves as a natural geographic unit for planning and management of water resources, the concept of the basin should be broad enough to encompass the engineering possibilities of integration and to address the commonalty of problems of the various river systems. In most cases the sub-basins would not constitute a unitary whole and sub-basinwise planning may not be desirable.

The river system

The main stem of the Brahmaputra is known as Tsangpo in Tibet, Siang or Dehang in the upstream area in India, and Brahmaputra in the rest of India and Bangladesh. The approximate distribution of the length and catchment area is given in Table 4.1. The average annual flow of the Brahmaputra at some salient points is given in Table 4.2.

The Barak-Meghna subsystem consists of the Barak river originating in Myanmar, which flows to the state of Manipur in India; thereafter, the river flows through the Cachar plains into Bangladesh. The distribution

Table 4.3 Countrywise length and catchment area of the Barak

Country	Length (km)	Catchment (km^2)	Average annual yield (billion m^3)
Myanmar	–	752	–
India	564	26,193	29.6 (Badarpurghat)
Bangladesh	336	–	–

of the length, catchment area, and water resources of this subsystem is given in Table 4.3.

The broad development objectives

As stated earlier, the region has a predominantly rural agricultural population. Providing reasonable access to clean water for sustenance of life should be a high-priority objective. As economic development of the area would be closely linked with agricultural income, water use in the agricultural sector also requires considerable priority. Navigation is essential for sustaining the riverine ports and harbours in the area; also inland navigation would have a large role to play in opening up the hinterland for economic integration, and so the navigational demands for water require a comparatively higher priority than in the rest of the South Asian region. The area has considerable industrial potential on account of the availability of various minerals. There is also a large potential for developing groundwater resources. In this context, the unusually high hydroelectric potential available in the Brahmaputra-Meghna basin seems to hold the key to industrial, economic, and agricultural development of the sub-basin.

However, various environmental considerations and constraints – such as the need to provide for proper resettlement and rehabilitation of the people affected by storage reservoirs, the need for conjunctive use of surface water and groundwater to maintain groundwater levels within a desirable band, the need for maintaining the ecology of the riverine fish life, and the need to maintain the salinity levels and sedimentation pattern in the rivers – are also important in planning water development.

Both the Brahmaputra and the Barak-Meghna sub-basin suffer from severe flood and erosion problems. The floods of the Brahmaputra and Barak are an annual feature, although some 3,600 km of embankments have been built on the main Brahmaputra river and its tributaries and another 800 km on the Barak. The average annual flood damage amounts to around Rs. 450 million (1977 price level; US$1 = Rs. 44) annually in the north-east region of India, mostly in the Brahmaputra-Barak area. In the alluvial plains the river flows in a braided pattern, and because of the shifting braids, an area of land with a width of about 10–15 km is more or less unavailable for any other use. Considerable damage and loss of land by erosion is caused even where the land is much above the normal high-flood levels. There are possibilities of both flood control through long-term structural measures like storage reservoirs and further erosion control through large river protection works, including the possibilities of changing the braided nature of the river into a meandering nature. These possibilities of flood and erosion control are largely compatible with the need for navigation development.

Considering these aspects, the author would prioritize the sectoral water use within the two subsystems as firstly water supply, secondly agricultural use of water, including provision for flood control, and thirdly use of water for energy development and navigation. The use of water for environmental maintenance is not mentioned as a separate use, but needs to be considered as an essential constraint in planning and development.

In broad economic terms, economic efficiency would be the major but not the only consideration in planning water development. The problems of floods, erosion, and lack of transport seem to have acted as a considerable constraint in the economic development of the Brahmaputra and Barak areas in the past. It could be argued that this lack of development may have led to development in other areas. However, the redistribution of economic opportunities could be considered as an important goal to bring back these lost opportunities of the region. Similarly, possibilities of integrated water development would allow economic development of the comparatively water-poor areas of the Ganges-Brahmaputra-Meghna system. Such economic development would also allow more equitable availability of employment to the vast rural population of the region. Thus the redistribution objective and the consequent goals of greater flood and erosion control, more agricultural development through irrigation in the water-scarce areas of the Ganges subsystem, greater navigation potential, etc., should receive much importance in the planning and development of the Brahmaputra and Meghna.

Water demand in north-east India – 2050

Projection of population

The population in north-east India is growing faster than in the rest of the country. The overall annual growth rate from 1901 to 1991 was 2.2 per cent per annum; in the decade 1981 to 1991, it was 2.4 per cent per annum. By 1991 the population was about 4 per cent of the all-India total. Presuming this to become 5 per cent by 2050, and using the medium projection of India's 2050 population at 1.64 billion, the population of north-east India by 2050 is estimated at about 80 million. This has been checked by a graphical extrapolation.

The present rate of urban population as a percentage of total population in India is 26.1 per cent, and that for north-east India is 13.9 per cent. It is presumed that the ratio for north-east India would equal 26.1 per cent by 2050. On this basis, the north-east India population in 2050 is taken as 21 million urban and 59 million rural.

The estimated large and small livestock populations by 2050 are 35 million and 46 million respectively.

Food requirement

On the basis of present consumption of around 500 gm per capita per day, the 2050 requirement of food grain per capita is taken as 590 gm per day. The projected 2050 population of 80 million in north-east India would, at this rate, require 16.4 million tonnes of food grain annually. However, development of irrigation in this region of India is not encouraging, and the region is food deficient: production of food grain in 1991–92 amounted to only 5.14 million tonnes. Unless adequate attention is paid to increasing food production, the region will have to depend largely on imports. However, considering the large water potential and available land, self-sufficiency is a desirable goal.

A plausible plan for irrigation development

The entire north-east region has a total cultivable area of 6.28 million hectares. The present gross cropped area is 5.29 million ha and the net cropped area is 3.729 million ha. Gross irrigation is practised presently in only 0.93 million ha. The yields are very poor: the yield of paddy in this region is only about 1.3 tonnes per hectare as against an achievable yield of 3–4 tonnes per hectare.

According to the report of the National Commission on Agriculture (1976), the scope of irrigation envisaged in Assam by 2025 is 2.5 million ha. The net irrigable area of Assam has been assessed as 3.27 million ha and that of the entire north-east region as 3.7 million ha. It is assumed that, by 2050, the entire net irrigable area of the north-east region may have to be covered by irrigation.

An estimation of potential irrigation development and food production is presented in Table 4.4.

Table 4.4 Plan for irrigation and food production in north-east India in 2050

Category	Present status	Probable scenario in 2050
	(million ha)	(million ha)
Geographical area	25.50	25.50
Cultivable area	6.30	6.30
Net sown area	3.70	4.70
Gross sown area	5.30	7.20
Area sown more than once	–	2.50
Gross irrigated area		
Surface source	0.95	2.30
Groundwater source	0.21	1.40
Gross area under rainfed crop	4.10	3.50
Irrigated area under food crop	–	3.30
Irrigated area under other crops	–	0.40
Rainfed area under food crop	–	3.20
Rainfed area under other crops	–	0.30
	(million tonnes)	(million tonnes)
Food production in equivalent Paddy		
Irrigated land @ 3.5 t/ha	–	11.55
Unirrigated land @ 1.6 t/ha		5.10
Total food production	–	16.65

Table 4.5 Estimated water demands

Category	Per capita requirement per day in litres (lpcd) (gross withdrawal)	
	1990	**2050**
Rural population	40	135
Small livestock	4	7
Large livestock	25	45
Urban population	90	200

Estimation of gross water demands

The rates of demand assumed for the rural and urban population, as well as small and large livestock, are given in Table 4.5. At these rates, the domestic and livestock demands are as given in Table 4.6.

For a preliminary assessment of industrial demand, it is assumed that the gross (withdrawal) requirement for industrial use would equal domestic (urban plus rural plus livestock) demand. For agricultural purposes, the average annual depth of irrigation water (gross) required for various crops is estimated as 1.6 metres for surface water sources and 1.2 metres for groundwater sources. An assessment of the monthly require-

Table 4.6 Gross domestic water requirement in 2050

Category	Population (million)	Water demand (lpcd)	Gross water requirement (million m^3/d)
Rural	59	135	8.0
Urban	21	200	4.2
Small livestock	46	7	0.3
Large livestock	35	45	1.6
Total daily requirement			14.1
Total annual requirement (gross)			5.147 bcm

Table 4.7 Projected gross irrigation water requirement for north-east India, 2050

	Water requirement (bcm)						
	North bank		South bank		Total		Total
Month	Surface	Ground	Surface	Ground	Ground	Surface	(Ground + surface)
February	0.20	0.10	0.30	0.10	0.20	0.50	0.70
March	0.20	0.80	1.90	0.90	1.70	2.10	3.80
April	1.90	0.80	2.60	1.20	2.00	4.50	6.50
May	1.50	0.70	2.00	0.90	1.60	3.50	5.10
June							
July	0.80	0.40	1.00	0.50	0.90	1.80	2.70
August	0.90	0.40	1.20	0.50	0.90	2.10	3.00
September	1.70	0.80	2.30	1.10	1.90	4.00	5.90
October	1.10	0.50	1.50	0.70	1.20	2.60	3.80
November	1.50	0.70	2.00	0.90	1.60	3.50	5.10
December	1.40	0.60	1.90	0.90	1.50	3.30	4.80
January	1.30	0.60	2.20	1.00	1.60	3.50	5.10
Total	12.50	6.40	18.90	8.70	15.10	31.40	46.50

ment for various crops is presented in Table 4.7: the requirement of 46.50 billion cubic metres (bcm) is for an irrigable area of 3.30 million ha in Assam, a state in north-east India. On a pro-rata basis, the total requirement for 3.70 million ha of irrigable area in the north-east region of India would be about 52.15 bcm (35.21 bcm from surface water and 16.93 bcm from groundwater).

Estimation of evapotranspiration needs of all uses

A large portion of domestic, industrial, and irrigation water is available as return flow; by making certain assumptions, it is possible to estimate consumption.

For rural use, including livestock, it is assumed that 50 per cent of the withdrawal would evaporate from land irrigation, use in kitchen gardens, etc., or otherwise as uncontrolled evapotranspiration, and the rest would return to the river or be retained as groundwater. For urban domestic use, it is assumed that 30 per cent would be consumed in parks, gardens, water coolers, etc., and the rest would return to the river or be retained as groundwater. For industrial use, it is assumed that 20 per cent would be consumed and the rest would return to the river or be retained as groundwater.

For agricultural use (irrigation), the depth for surface irrigation has been assumed as 1.6 metres (gross, per head), on the presumption of paddy being the principal crop. The general distribution of this would be evapotranspiration 0.5 metres, deep percolation through fields 0.3 metres, losses from field channels 0.2 metres, and canal losses 0.6 metres. Thus, of the delta of 1.6 metres, only 0.5 metres will be used for crop evapotranspiration and 1.1 metres would be available for reuse/return through surface water and groundwater. However, it is presumed that of this 1.1 metres, another 0.2 metres is lost as uncontrolled evapotranspiration from swamps, etc. Thus the net evapotranspiration requirement would be 0.7 metres – 44 per cent of the gross requirement.

For groundwater the delta being used is 1.2 metres, assuming that distribution (canal) losses from groundwater would be very small at 0.2 metres. Again, assuming the additional swampy evaporation to be low, at 0.1 metres, the net evapotranspirational needs would 0.6 metres – 50 per cent of the gross requirement.

Total gross and net water demands

Total water requirement by 2050 for meeting domestic, industrial, livestock, and agricultural demands is presented in Table 4.8.

Non-consumptive in-situ requirements

The approximate navigational requirements would be 400 m^3/sec on the Barak (draft exceeding five metres post storage), and 2,000 m^3/sec on the Brahmaputra (draft of two to five metres).

Resource availability

Water resources of India

According to one assessment (CWC 1993), the average annual flow in the river systems of India is estimated to be 1,869 bcm. Replenish-

Table 4.8 Total gross and net water demands

Sector	Gross demand (bcm)	% of consumption	Net demand (bcm)
Domestic water supply			
Rural			
Domestic	2.920		
Livestock	0.694		
Total (rural)	3.614	50	1.807
Urban	1.533	30	0.459
Sub-total (1) (water supply)	5.147		2.266
Industrial (2)	5.147	20	1.060
Agricultural			
Surface water irrigation	35.200	44	15.500
Groundwater	16.900	50	8.500
Subtotal (3) (agriculture)	52.100		24.300
Total (1) + (2) + (3)	62.39		27.630

Table 4.9 Water resources availability in Bay of Bengal basin

River sub-basin	Average annual water potential (km^3/annum)	Per capita flow availability (m^3/annum)
Ganges-Brahmaputra-Meghna	1,110.62	18,061
Ganges	525.02	1,471
Brahmaputra	537.20	18,417
Barak	48.40	7,646
Minor rivers draining into Bangladesh and Myanmar	31.00	14,623
Godavari	110.54	2,048
Krishna	78.12	1,248
Cauvery	21.36	728
Subernarekha	12.37	1,307
Brahmani-Baitarani	28.48	2,915
Mahanadi	66.88	2,513
Pennar	6.32	651

able groundwater resources have been assessed as about 452 km^3/annum. The water resources potential for some of the important river basins in the geographical Bay of Bengal basin is given in Table 4.9.

Table 4.10 Water resources potential in north-east India

Basin	Average availability (km^3/annum)	Availability per ha of cultivable area (m^3/ha/annum)	Utilizable Surface water (m^3/ha/annum)	Utilizable Ground-water (m^3/ha/annum)
Brahmaputra	537.2	44,232	24	27.9
Barak	48.4	43,447	–	1.8
Other rivers in Tripura, Manipur Mizoram and Nagaland	31.0	–	–	1.53

Water resources in north-east India

The total water resources potential of the river systems in the north-east region of India and per capita average annual availability as assessed is given in Table 4.10.

Storage possibilities

Since most of the precipitation of the region is received in the summer monsoon (April to October), storage development is essential for both conservation and flood control. In the Brahmaputra-Barak areas, which contribute about 32 per cent of India's surface water resources, the likely ultimate storage capacity is only about 72 bcm – only 12 per cent of average annual flows. Although storage capacity in the region is only about 19 per cent of the country's ultimate storage capacity, it has about 37 per cent of the country's estimated hydropower potential besides the potential to moderate floods in the Brahmaputra and Barak valleys to a great extent on a long-term basis.

One of the recommendations of the fourth national power plan formulated by the Central Electricity Authority in March 1997 is to accelerate the pace of hydropower development to add 58,000 MW of hydro capacity in the next 15 years. With this objective in view, hydropower development strategies in the north-east region need to be framed, to provide at least 50 per cent of this projected 58,000 MW of hydro development by 2030.

Development of all available storage sites in this region of India may not be practicable because of social constraints in the resettlement of populations and possible technical constraints. Keeping in view the huge power potential and excluding the storage sites having serious social constraints, the storage facilities given in Table 4.11 have been identified in the region and may be developed in a phased manner up to 2030.

Table 4.11 Storage sites identified for development

Basin/sub-basin	Live storage (bcm)	Installed capacity (MW)
By 2010		
Subansiri at Gerukamukh	0.64	600
Tipaimukh	9.00	1,500
Bairabi	1.78	75
Jadukata	0.72	450
Siang (lower)	0.80	2,000
Noa-dihing	0.13	75
Kulsi	0.69	36
Someswari	0.09	130
Um-N-got	0.03	710
Total	13.88	5,579
By 2020		
Subansiri (upstream reservoirs)	6.02	6,700
Lohit	2.75	3,000
Kameng	1.07	1,100
Kulsi	0.69	36
Total	10.53	10,800
By 2030		
Siang (Yinkiong)	5.30	10,700
Siyom (Kaying)	0.50	700
Debang	2.00	1,000
Total	7.80	12,400

It is thus proposed to develop about 32.21 bcm of storage capacity in the region by 2030, which is abut 40 per cent of the ultimate storage potential of the region.

Engineering possibilities

Diversion

In the Indian context, 'run-of-the-river' diversions were developed first, up to 1950, and storage development followed, in order to augment water use. Major groundwater development also took place, notably after 1970, the timing being dictated perhaps more by the spread of rural electrification than by the exhaustion of the diversion possibilities. The relevance of this development pattern to north-east India needs examination.

The dependable flow of the Brahmaputra and Barak in the lean-flow

Table 4.12 Lean-season water requirement in north-east India

Use	February surface irrigation (million m^3)	Gross (m^3/s)	Net (m^3/s)	March surface irrigation (million m^3)	Gross (m^3/s)	Net (m^3/s)
Surface water irrigation	500	207	93	2,100	784	345
Groundwater irrigation	200	83	42	1,700	635	318
Water supply (at uniform annual rate)	429	166	73	429	166	73
Industrial water supply (at uniform annual rate)	429	166	33	429	166	33
Total		622	239		1,751	769

period would be of the order of 3,000 m^3/sec and 45 m^3/sec respectively at their exit points. The total groundwater potential of the two sub-basins, at about 31 bcm per year, can support for 240 days/annum a draft of about 1,500 m^3/sec. From a simple hydrologic point of view, the groundwater draft may in the long run lead to more reduction in the surface flows. But together from both sources, about 3,000 m^3/sec of water is available. Against this, the conservation requirements from these systems for two sample months are as given in Table 4.12.

Thus the net withdrawals from the system, including groundwater, would be of the order of 239 m^3/sec and 769 m^3/sec in February and March respectively, which is lower than the lean flow of 304 m^3/sec. The actual effect of the withdrawal could be even less, as much return flow from the withdrawals in other months would also be flowing during the lean season. But even on a conservative basis, a reduced flow of the order of 2,500 m^3/sec would be available in the river systems after all conservational use. The non-consumptive requirements, including the navigational requirements, have been discussed earlier. This estimate indicates that this low flow would be sufficient.

Therefore, if self-sufficiency in food grain in north-east India is to be attained, providing essential water supply and maintaining a limited navigational draft in the Brahmaputra and Barak are the only goals of development, and if electrical energy for groundwater development could be found from other sources, storage development is not essential. However, in the opinion of the author this would preclude many attractive development options. Storage-based hydropower development would be essential for both agricultural and industrial development. The long-term flood control possible through storage facilities would shift economic activities to the flood-prone areas. These developments would require a much better transport network and riverine navigation with improved draft, which is possible with storage facilities. Also, the storage sites would facilitate a more equitable availability of water in the three subsystems of the Ganges-Brahmaputra-Meghna basins.

The engineering possibilities for run-of-the-river development would consist of a series of weirs or barrages, which can be built at the places where tributaries of the Brahmaputra emerge out of the hills, and similar barrages on the Barak. Such structures have already been built on the Ganges and its tributaries. In the Brahmaputra basin, such barrages have been erected on the Dhansiri, Jamuna (a tributary in the Kopili Kolong basin), Shukla, Patrodisa, Hawaipur, Longa, Rajadubi (lift), Dikhari, etc. More schemes, like the Baralia and Bordikrai, are under construction. These schemes are undoubtedly small, and their construction and maintenance is sometimes problematic due to sedimentation, shifting courses, etc. However, they demonstrate the possibilities. Similarly, groundwater development is possible almost throughout the plains of the Brahmaputra and Barak valleys, provided electrical energy is available.

Storage development

As stated earlier, storage development of the order 30 bcm (live) is possible in the Brahmaputra and Barak sub-basins. An attempt is made to evaluate approximately how much additional water would be made available through these. It is assumed that 20 per cent of the storage space is reserved as an empty space to hold flood waters for a period of about 10 days during the peak river floods, by proper integrated operations. This would allow the cutting off of a peak of 15,000 m^3/sec, giving a small but significant flood relief. By proper rule-curve operation and use of joint flood conservational storage, a conservation storage of about 24 bcm would be available (lake evaporation, etc., being offset by the joint use). This could provide, in the three-month lean-flow period, an additional flow augmentation of the order of 3,100 m^3/sec. Again, by proper integrated operations, the augmentation of the lowest 10 daily flows by about 4,000 m^3/sec appears practicable. This additional low flow can then be utilized for both additional navigational requirements in the Brahmaputra and Barak and augmentation of lean flows in the Ganges.

Flood management, drainage, and erosion

The north-east region of India, particularly the Brahmaputra and Barak valleys, is subjected to frequent and damaging floods almost every year,

Table 4.13 Land use in the Brahmaputra valley

Category	Area (ha)	%
Land	7,089	100.00
Forest cover	1,506	21.25
Agricultural land	3,917	55.23
Built-up area	18	0.26
Water bodies	551	7.79
Wasteland	127	1.80
Other	970	13.67

Table 4.14 Flood damage in Assam

	Maximum	Average
Total area affected (million ha)	3.820 (1988)	0.96
Crop area affected (million ha)	1.440 (1988)	0.23
Total damage to crops, houses and public utilities (million rupees)	6,638.4 (1988)	810
Lives lost (numbers)	226 (1988)	39
Population affected (in million)	12.46 (1988)	2.43

with the Brahmaputra becoming one wide sheet of water covering a large area during the monsoon. The monsoon flood problem in the Barak valley is equally severe. The land-use pattern of the Brahmaputra valley is given in Table 4.13.

The problem of floods in the flood plains of the Brahmaputra can be summarized as bank spilling, drainage congestion due to adverse outlet conditions, and morphologic problems caused by heavy sediment load related to land use and seismic activities. Flood damage statistics have been collected in India since 1953. Table 4.14 gives a picture of the maximum and average annual flood damage in Assam during the period 1953–1994, as compiled by the Central Water Commission of India. In addition to this damage, there is considerable indirect loss due to disruption of rail and road communication, lower crop production, and extra expenditure on relief and rehabilitation, etc.

No storage reservoirs have yet been created in north-east India in the Brahmaputra and Meghna river system for flood control measures. Measures taken so far are embankments, drainage channels, town protection, anti-erosion works, and sluices. Physical achievements up to 1988 in the Brahmaputra-Barak area are shown in Table 4.15.

During less severe floods the embankments, which are normally designed for a 25-year flood, have afforded some degree of protection, but during high floods they are often subjected to breaches, causing severe damage.

Table 4.15 Flood and erosion control measures in north-east India

Description	Brahmaputra valley	Barak valley	Total
Embankments (km)			4,448
(a) Brahmaputra/Barak	1,013	251	
(b) tributaries	2,634	550	
Anti-erosion schemes, including town protection works (km)	431	112	533
Drainage channels (km)	599	247	846
Sluices (number)	56	29	85

Measures suggested for flood management

Storage reservoirs

In India, single-purpose flood control reservoirs are normally not found economical, but multipurpose reservoirs often have flood control as an important objective.

The five major tributaries of the Brahmaputra, namely the Dehang, Jia-bharali (Kameng), Debang, Lohit, and Subansiri, together contribute about 70 per cent of the average annual yield of the Brahmaputra at Pandu. Storage reservoirs on these five rivers and on the Barak at Tipai-mukh are under various stages of investigation, and can generate 28,500 MW of cheap eco-friendly peak hydropower besides moderating floods to a great extent.

Embankments

As stated, long embankments have already been constructed, and few additions are planned.

Drainage schemes

When storage reservoirs are constructed on some of the major tributaries and the flood peaks are brought down, most of the drainage-congested areas would automatically get cleared. However, to provide some interim solutions to minimize the difficulties created by drainage congestion, identified problem areas are to be studied and suitable remedial measures are to be taken up.

Anti-erosion measures

The Brahmaputra has braided channels in most of its traverse through the plains. One of the most important areas affected by erosion, amongst others, is Majuli, the largest river island in the world, on the north bank of the Brahmaputra. A number of erosion control measures are planned.

Table 4.16 Distribution of hydropower potential by tributary

River system	Potential at 60% lean flow (MW)
Upper Brahmaputra	789
Lohit	4,152
Dihang-Dibang	13,615
Subansiri	6,893
Kameng	1,982
Kalang (Kopili)	518
Barak and neighbouring rivers	3,908
Total (north-east India)	31,857
Tista, Jaldhaka	3,063
Total (Brahmaputra and Barak)	34,920

Non-structural measures

A flood warning system is already working effectively in the area. Flood-plain zoning, although desirable, has not been put into practice. Watershed management and discouragement of *jhooming* (shifting cultivation in hilly areas) is receiving much attention, particularly from an institute of the All-India Council of Agricultural Research established for the purpose. The present soil loss of 40 tonnes per hectare in the area under *jhoom* needs to be reduced. Socially acceptable alternative packages need to be developed and implemented.

Hydropower development

Hydropower potential

Out of India's total hydropower potential of 84,044 MW (at 60 per cent of lean flow), about 59,160 MW is accounted for by the Himalayan river basins of the Indus, Ganges, and Brahmaputra, and about another 5,250 MW for the Barak and other southern tributaries of the basin. The north-east Himalayan rivers constitute about 55 per cent of the total Himalayan hydropower potential. According to a reassessment of hydropower potential in India carried out by the Central Electricity Authority, 226 schemes have been identified in the Brahmaputra and Barak river systems with a total hydro potential of 34,920 MW. The distribution by tributary is given in Table 4.16.

Present power scenario

The total electrical energy consumption in north-east India was 3,736 million units in 1995–96, which was only about 1.33 per cent of the total energy consumption in the country. The per capita consumption in the region is of the order of 100 units, against all-India average of about 333 units for 1995–96.

The power supply position in north-east India changed from energy surplus (2.4 per cent) in 1984–85 to a deficit (–4.6 per cent) in 1990–91. According to the fourth national power plan prepared by the Central Electricity Authority in March 1997, the peak local demand of the region would increase from 1,313 MW (1997) to 3,661 MW (2011). Similarly, the energy demand would increase from 5,855 GWh (1997) to 17,553 GWh (2011).

Although the present installed capacity of about 1,628.95 MW is greater than the peak demand during 1996–97, the existing generating stations are not able to meet peak demand for various reasons, including the very low plant load factor of thermal units.

Although the present demand pattern may not justify setting up big hydropower projects, this would have to be considered in the wider context of accelerated economic development and transmission of cheap power to the power-starved regions of India and neighbouring countries.

Power for economic development

It is obvious that with the current pace of industrial development in the region, the development gap between the north-east region and the rest of India will continue to increase. To bridge the gap it would be necessary to plan for a much faster rate of development of industries and the agricultural sector. The irrigation development plan presented earlier in this chapter already requires large-scale pumping of groundwater. Both industrial development and groundwater development would require corresponding power development.

With the availability of cheap and assured power, it is likely that per capita income and rural electrification can grow at a faster rate. Every household in villages in north-east India, be they on the plains or in the hills, has at least one handloom unit each. Adopting composite weaving, reeling, and yearning farming and converting the handlooms to powerlooms would give the rural economy a fillip.

The employment impact of the anticipated development programme would be significant. Currently, the employment-income ratio in the north-east region is higher than the all-India average. It is anticipated that with the development of various sectors of the economy in the region with judicious use of power, employment in the organized sector will rise to almost five times the present level within the next 25 years.

Table 4.17 Likely power transfers from north-east India

Planning scenario date	Installed capacity in north-east (MW)	Peaking capacity in north-east (MW)	Peak demand: North-east region (MW)	Peak demand: All India (MW)	Power evacuation (MW)
2010	7,593	5,698	3,200	170,000	2,495
2020	16,193	12,145	5,700	270,000	6,445
2030	28,893	21,670	8,200	355,000	13,470
2040	59,256	44,442	10,700	450,000	33,742

Note: Peaking capacity is assumed to be 75 per cent of installed capacity.

Development strategies for hydropower

Two of the important conclusions and recommendations of the fourth national power plan developed by the Central Electricity Authority in March 1997 are as follows.

- Additional capacity of over 150,000 MW is required to meet the 176,647 MW peak demand and 1,058 billion units of energy requirement by the end of 2012.
- The pace of hydro development must be accelerated to add 58,000 MW of hydro capacity over the next 15 years. In a low hydro development scenario, additional thermal capacity of 9,000 MW would be required. Effective measures to attract private investment within the country and outside will be required.

Much of the additional 58,000 MW of hydropower may have to come from the north-east region of India, particularly since the average generation cost in the region would be low and the population likely to be affected by the projects would also be low.

Transfer of power from north-east India

In view of the likely severe power shortages in other regions, which are likely to grow further in the years to come, the cheap hydropower of the north-east region can be beneficially transferred to the other regions of India and perhaps the neighbouring countries. Thus setting up large hydropower stations in the region within the next 10 years is a feasible proposition even if growth of demand for power in the region is moderate.

Table 4.17 shows the likely installed capacity in the north-east region, likely peak demand in the region, likely demand in India, and power likely to be evacuated to other regions of India in different planning scenarios.

Financial aspects

Development of the huge hydropower resources of north-east India will call for huge investment. Owing to the financial constraints in the government sector, the available resources may not be adequate to meet the envisaged power and allied sector development programme. Thus a policy needs to be evolved, with special consideration of the geographical and other constraints of the north-east region, for financing the megaprojects with private-sector funding. This could involve joint ventures or international participation for improving the health of the power and allied sectors and accelerating economic development.

Navigational development

Present scenario

At present the Department of Inland Water Transport (DIWT), a government organization, is operating 42 ferry services on the Brahmaputra and 24 ferry services on the Barak. With this ferry network, Assam carries nearly 60,000 passengers and 2,000 tonnes of freight every day. In addition to the people of Assam, the DIWT ferry operation has benefited the people from other north-eastern states of India, as some of the major ferry services connect Assam with the neighbouring states. The cargo-handling capacity of the commercial fleets, with seven tugs and 25 barges, is currently 11,900 tonnes. In 1996–97 the DIWT could carry only 2,117 tonnes. The region has the largest amount of navigable waterways in India, with a length of 3,844 km.

The national transport policy of India indicates that for bulk cargo like coal and fertilizers, navigational costs are substantially lower than road transport costs.

Development strategy

Navigational requirements in the region are considerable, particularly on the Brahmaputra. The present problem of increasing the navigational draft in the Brahmaputra seems related to the braided nature of the channel. The numerous braided channels not only occupy a considerable width of the valley, but also divide the low flow so that a defined and stable navigational channel is not available.

In the short term, local river control measures, including annual measures like *bandalling* (a method of seasonal maintenance of low-flow navigational channels by inducing sedimentation elsewhere througn bamboo bundles), if properly planned can much improve the navigability, or for the same level of draft reduce the flow requirement.

In the long term, there seems to be scope for jacketing the river at some points to convert, over a period of time, the river regime from that of a braided stream to that of a meandering stream mostly flowing through a single channel. Such transformations have been done for rivers like the Rhine, and there is a need to study the possibilities for the Brahmaputra on elaborate mathematical models depicting the flows and sediment transport. If this can be done, it would have the further advantages of improving navigability and increasing availability of land for agriculture and other uses.

Structural measures, such as provision of a series of barrages with navigational docks for increasing depths for navigation at lower flows, could be thought of. However, these may become practicable only if navigational requirements build up in a large way.

The Brahmaputra development strategy

The development strategy for the Brahmaputra and Barak proposed in this chapter, on the basis of the facts and discussions presented earlier, is as follows.

- There is a need for a holistic look at the needs of the Ganges-Brahmaputra-Meghna basin.
- The water-use plan should meet the following priority needs:
  - needs for domestic water supply (rural and urban) in the Brahmaputra-Meghna region;
  - needs of agriculture and flood control systems in the Brahmaputra-Meghna region;
  - industrial needs and needs for navigational improvement in this region;
  - needs for equitable distribution of water resources for all uses in the Ganges-Brahmaputra-Meghna basin;
  - needs for possible equitable distribution of water in a wider region, such as the geographical Bay of Bengal basin.
- With regard to the agricultural needs, the overall objective could be self-sufficiency in food grain in north-east India, combined with full utilization of land resources.
- In planning irrigation in north-east India, comparatively much greater stress on groundwater utilization is necessary. For this purpose, energy would have to be made available through hydropower development.
- Extending flood and erosion control is a priority need, so that eco-

nomic activities are attracted to the area and regional redistribution of income and employment is encouraged.

- An approximate estimation of water demands for various uses in the Brahmaputra-Barak areas of India has been made. Even while this estimate needs to be improved and expanded to include the downstream areas, some general results seem to emerge.
  - The water needs for domestic supply and agriculture are relatively small compared to the runoff of the river. This fact, combined with the large availability of groundwater, indicates a considerable possibility for achieving self-sufficiency in food grain even without large storage-based surface irrigation systems.
  - However, for adequate availability of electrical energy, for both agriculture and industry, hydro-electric storage facilities would be necessary. Similarly, for providing long-term flood control, storage development is necessary.
  - A hydro-electric potential of the order of 31,857 MW at 60 per cent of low flow is available. This will be enough to meet the regional demands and would provide considerable scope for earning revenue.
  - Significant flood control in the rivers can be achieved through integrated operation of a part of the storage space, to be reserved for the purpose.
  - The navigational demands for water are important. There is a considerable scope for improving navigation, even with lean low flows, by proper river training works.
- There is a scope for changing the regime of the Brahmaputra from a braided pattern to a meandering pattern through a long-term plan of large-scale river training. Such a plan, if implemented, will improve navigational conditions and also make more land available for agricultural and other purposes. Such possibilities require elaborate study through modelling.
- There is also scope, in the long term, for providing greater navigational depths on the Brahmaputra during smaller low flows through the construction of a series of barrages and locks. However, for the present such possibilities need not be considered because the large likely cost is not commensurate with navigational demands, and salinity control and other ecological considerations would require maintenance of a flow in the channel.
- A large augmentation of river low flows of the order of 4,000 cumecs would be possible through storage development.
- Even while the first priority for the augmented flow would be to meet the needs of the region (water supply, agriculture, navigational improvement, etc.), a substantial portion of the augmented flow would be available for priority uses elsewhere in the Ganges-Brahmaputra-Meghna basin.

- The Ganges-Brahmaputra-Meghna basin has considerable scope for water-use integration through interconnecting canals. Thus, the holistic development of the three subsystems through the integrated operation of storages and links is practicable. Such a plan would improve agriculture throughout the basin, provide greater navigational facilities in all the rivers, and allow generation of a large block of hydro-electric energy.

REFERENCES

CWC. 1993. *Reassessment of Water Resources Potential in India*. Publication No. 6/93, March. New Delhi: Central Water Commission.

National Commission on Agriculture. 1976. *Final Report to the Government of India*. New Delhi: Ministry of Agriculture.

5

Development and management of water resources in Bangladesh: Post-1996 treaty opportunities

Ainun Nishat

Introduction

The Ganges waters treaty, signed between India and Bangladesh in December 1996, has opened the doors of opportunity for development and management of water resources in the Ganges-Brahmaputra-Meghna (GBM) region in an integrated way. The national framework and plans will provide a basis for regional approaches and considerations. The treaty provides a framework for sharing the waters of the Ganges during the dry months and calls for sharing arrangements to be agreed for all other common rivers. Bangladesh can now plan, harness, and develop the water resources of the Ganges.

Bangladesh is situated in the deltaic part of the region through which three mighty rivers and their tributaries and distributaries flow, namely the Ganges, the Brahmaputra, and the Meghna. A huge volume of water enters the country from outside and flows into the Bay of Bengal. In fact, more water enters Bangladesh each year as surface inflow that can ever be used effectively – enough to submerge the country under more than nine metres of water, of which only 7 per cent originates from within the country. On average, during each monsoon season from June to September, one-third of Bangladesh is stricken by flood calamities and two-thirds of the country is vulnerable to flood. This apparent surfeit of water conceals important seasonal and spatial water shortages that prevail during the winter, which generate drought or drought-like conditions.

During the dry months (November to May), flows of major rivers decrease drastically. Fresh water becomes scarce for use in agriculture, fisheries, navigation, industries, drinking, or domestic purposes.

This annual pattern of too much water during the monsoon and too little during the dry season dominates life in Bangladesh. The cycle of floods due to overabundance through to droughts from dire water scarcity brings further suffering in the form of land salinization, siltation, and river instability. Bangladesh has unilaterally sought to reduce the severity of flooding, facilitate river transport, and improve irrigation, but national efforts alone cannot resolve problems in a multinational river basin. For the well-being of the hungry people in the region, it is essential that stream flow regulation, flood control, and water resources development should proceed in an integrated regional fashion. The prevailing wisdom is that investment in large surface water projects would reduce the element of uncertainty in the availability of water in the dry season as well as attenuate the flood levels. Another major benefit would be generation of hydropower, which would go a long way towards keeping the production of greenhouse gases to a much lower level.

As there was no effective water-sharing agreements for these international rivers, nor any efforts to develop water resources in an integrated and comprehensive manner, Bangladesh had to remain satisfied with what it could do within its own borders. No major surface water development project could be taken up, and planning scenarios were unnecessarily restricted to harnessing and utilization of groundwater.

This chapter gives a historical background to the 1996 treaty, and makes an analysis of the provisions and functioning the treaty in its first year of operation in the dry months of 1997. There is then a discussion of the need for a control structure over the Ganges, and opportunities and issues for other common rivers. Possible linkages with regional development opportunities such as navigation, hydropower generation, watershed management, and fish movement are also highlighted. In the concluding section the existing institutional framework and the role of the Joint Rivers Commission are evaluated.

The history of water disputes between Bangladesh and India

Dispute over the Ganges flow was recognized in 1951, when Pakistan protested to India about the proposed Farakka barrage. Negotiations have continued between India and Bangladesh (and prior to 1971, Pakistan) on this issue for the last 45 years. There were several short-term sharing agreements before the long-term treaty was signed in 1996. Several studies have analysed the background and history of the dispute and reviewed the whole process; notable among them are the works of Abbas (1982), Begum (1987), Verghese and Iyer (1993), Crow (1995), and Nishat (1996). A brief review is given here to identify and recognize the some of the major issues, which need to be resolved before further co-operation in the GBM region can be ensured.

Crow (1995) divided the history into several phases based on political considerations. On the basis of approaches and progress at negotiations, the entire process may be into clustered six phases (Nishat 1996).

Phase I: 1951–1974

Focus was on finalization of respective shares of the Ganges waters before the barrage was commissioned. Discussions centred mostly on respective claims and their justifications. The Bangladesh side staked claim over the entire flow and expected that reservoirs at upstream locations should be able to meet India's requirements. They also contested the technical merit of the stated objective of the Farakka barrage, which was to save the port of Calcutta. India, on the other hand, took a position that Bangladesh may need only a small quantity of water and the water was presently flowing wasted into the Bay of Bengal. The Indian side would ensure that the respective shares were finalized before commissioning of the barrage. In 1972, the Joint Rivers Commission (JRC) was set up to facilitate development of waters of the common rivers.

Phase II: 1974–1976

The issue of augmentation of flow was brought in. Bangladesh proposed conservation of huge monsoon flows in India and Nepal; and India proposed diversion of a large volume of the flows of the Brahmaputra into the Ganges through a link canal. Negotiation failed to arrive at an agreed formula for sharing. Operation of the barrage began in 1975, with concurrence from Bangladesh. An interim arrangement was made for a test run of a flow of 11,000 to 16,000 cusecs through the diversion channel for a period of 41 days only, releasing the remainder for Bangladesh. Without a renewal of the arrangement, in 1976 India withdrew waters unilaterally, causing disastrous effects in the Ganges-dependent south-west region in Bangladesh. This situation was repeated in 1977. The issue was taken to the United Nations by Bangladesh, and the General Assembly adopted a consensus statement according to which India and Bangladesh decided to meet at Dhaka at ministerial level for negotiations with a view to arriving at a fair and expeditious settlement.

Phase III: 1977–1982

After negotiations, an agreement was signed on 5 November 1977 for sharing the Ganges water for a period of five years during the dry months between January and May. The agreement provided for 34,500 cusecs for Bangladesh and 20,500 cusecs for Calcutta port during the leanest 10-day period (21–30 April). The agreement specified a schedule of sharing the flow on a 10-day basis; a guaranteed minimum of 80 per cent of the quantum mentioned in the schedule to protect Bangladesh against withdrawals in upstream reaches; a joint committee comprising representatives of both the governments to implement sharing arrangements; and carrying out investigations leading to augmentation of the Ganges flow at Farakka within three years. During this period Bangladesh and India exchanged their proposals for augmentation of the Ganges flow, and these were reviewed jointly by the parties. The Bangladesh proposal envisaged construction of storage dams in the upper reaches of the Ganges in India and Nepal to conserve the monsoon flows for augmenting the dry-season flows. The Indian proposal was for transferring waters of the Brahmaputra through a link canal across Bangladesh to the Ganges above Farakka. Neither proposal was accepted by the other party. On the plea that the obligation under the agreement had not been fulfilled, the treaty was not renewed despite there being provisions to do so, and it was allowed to expire in November 1982.

Phase IV: 1982–1988

The governments of Bangladesh and India signed a memorandum of understanding (MOU) in October 1982 for sharing the flows of the Ganges for another two dry seasons in 1983 and 1984. The MOU of 1982 included provisions for burden-sharing instead of the guaranteed minimum flow. Both parties were asked to carry out a pre-feasibility study of the schemes for augmentation proposed earlier. Accordingly, in 1983, both sides exchanged their respective updated proposals, but in view of the differences of opinion it was not possible to make any acceptable recommendation with regard to the optimum solution for augmentation of the dry-season flows of the Ganges at Farakka which could be urgently implemented. The MOU expired in 1984, and no formal arrangements for sharing the flows in future years were made.

Bangladesh made serious efforts to reach a formal agreement, but without success. In the absence of any agreement, no sharing was done during the dry season of 1985. After painstaking efforts the two countries signed another MOU in November 1985 for sharing the waters of the Ganges for three dry seasons commencing in 1986. Like the 1982 MOU, the MOU of 1985 included provision for burden-sharing instead of guaranteed flows. This time regional rivers were placed on the discussion table for arriving at sharing arrangements. Proposals on both sharing and augmentation of the Ganges flows were discussed. Both sides revised their proposals on augmentation and also rejected the other's schemes. A joint team of experts from India and Bangladesh visited Nepal for collection of data. Nepal showed keen interest in harnessing the waters as a joint programme between the three countries. The sharing arrangement of the Ganges flows according to the MOU of 1985 stopped in May 1988.

Phase V: 1988–1996

During this period negotiations for sharing continued, but without success. During a meeting the two heads of governments held in New Delhi in September 1988, the secretaries of water resources were assigned to work out an integrated formula for permanent/long-term sharing of flows of all the common rivers. In order to break the stalemate in the sharing arrangement, the Secretaries' Committee held six meetings alternatively in Dhaka and New Delhi over three years from April 1990 to February 1992. They emphasized the need for immediate allocation of the Ganges and Teesta waters on a priority basis, including sharing of other common river waters as mandated. The committee, however, could not find a general principle for sharing. The 1988 flood in Bangladesh created an institutional framework for cooperation in flood management. The relationship between sharing arrangements and augmentation proposals became a very critical issue. Bangladesh wanted to make progress in both aspects, but India made agreement to its augmentation proposal a precondition for sharing. India also continued to block any proposal for the involvement of Nepal in the discussions.

The prime ministers of the two countries agreed in 1992 that distress in Bangladesh need to alleviated immediately, but no water-sharing arrangements could be finalized. In 1995 several rounds of discussions were held at foreign secretary level.

Phase VI: 1996 to date

With changes in government in both the countries, negotiations were revived at the foreign secretary level and very soon the framework of a solution was agreed upon. After hectic activities a treaty was signed in December 1996 which went into operation from 1 January 1997. The treaty will remain in force for 30 years. Stages have been set out for sharing the water resources of all common rivers. The treaty also makes provision for augmentation of the flow of the Ganges. The flow arriving at Farakka was shared during the dry season of 1997. There were some minor problems in implementing the provisions of the treaty, which were resolved through consultations.

This historical background tells us that both sharing and augmentation must be resolved, but one should not be contingent upon the other. Now that the sharing issue has been resolved, serious efforts should be made towards solving the augmentation issue.

Brief features of the 1996 treaty

The full text of the 1996 Ganges water treaty is given in Appendix 1. The main features are:

- a more-or-less 50–50 sharing formula;
- three guaranteed 10-day periods for both countries when each will get 35,000 cusecs;
- the need for augmentation is recognized;
- the need for sharing all common rivers is recognized;
- principles of equity, fair play, and no harm to either party are recognized.

At the time of writing, the treaty had been in operation for two dry seasons. The flow arriving at Farakka and the share received by Bangladesh in 1997 and 1998 are given in Tables 5.1 and 5.2. One must take note that the flow being shared at Farakka is the residual flow after abstraction at upstream locations has taken place. Article II (ii) of the treaty states that the upper riparian will protect the flows at a specified level. This is a kind of guarantee to Bangladesh that there will be adequate flow at the point of sharing for meaningful operation of the treaty. Unfortunately, in the first year of operation the flow available for distribution was very low, which resulted in doubts being expressed about the sincerity of the upper riparian in honouring their commitments. However, the year ended with satisfaction for both the parties as the flow situation was brought within the expected range, and the flow received by the lower riparian in 1998 was satisfactory. As the flow arriving at Farakka was good, no difficulty was faced in implementing the provisions of the treaty.

Adverse impacts of reduction of the Ganges flow

The Ganges has been flowing through the territory of Bangladesh from time immemorial; the lives and livelihoods of its people, together with its flora and fauna, have been conditioned by the waters of this great river.

Table 5.1 Flows of the Ganges at Farakka and Hardinge bridge, 1997

Period		Actual flow reaching Farakka (cusecs)	Indian share as per formula given in Annexure-I of 1996 treaty (cusecs)	Bangladeshi share as per formula given in Annexure-I of 1996 treaty (cusecs)	Actual release to India (cusecs)	Actual release to Bangladesh (cusecs)	Observed flow of the Ganges at Hardinge bridge (cusecs)
January	1–10	1,02,180	40,000	62,180	40,161	62,019	70,829
	11–20	89,635	40,000	49,635	40,079	49,556	55,788
	21–31	88,672	40,000	48,672	39,788	48,884	50,045
February	1–10	85,604	40,000	45,604	40,000	45,604	48,430
	11–20	81,015	40,000	41,015	39,986	41,029	38,319
	21–28	77,399	40,000	37,399	39,012	38,387	25,689
March	1–10	66,170	33,085	33,085	32,681	33,489	23,291
	11–20	56,769	21,769	35,000	21,741	35,028	19,930
	21–31	48,487	35,000	13,487	31,959	16,528	13,823
April	1–10	50,481	15,481	35,000	20,344	30,137	17,857
	11–20	54,526	35,000	19,526	28,913	25,613	24,559
	21–30	63,933	28,933	35,000	28,868	35,065	27,695
May	1–10	66,728	35,000	31,728	35,006	31,722	26,578
	11–20	66,055	33,027	33,028	33,034	33,021	26,279
	21–31	63,309	31,655	31,654	31,666	31,643	27,520

Table 5.2 Flows of the Ganges at Farakka and Hardinge bridge, 1998

Period		Actual flow reaching Farakka (cusecs)	Indian share as per formula given in Annexure-I of 1996 treaty (cusecs)	Bangladeshi share as per formula given in Annexure-I of 1996 treaty (cusecs)	Actual release to India (cusecs)	Actual release to Bangladesh (cusecs)	Observed flow of the Ganges at Hardinge bridge (cusecs)
January	1–10	204,797	40,000	164,797	40,034	164,763	182,286
	11–20	175,566	40,000	135,566	39,975	135,591	154,292
	21–31	145,866	40,000	105,866	39,985	105,881	118,313
February	1–10	128,186	40,000	88,186	40,005	88,181	88,363
	11–20	101,841	40,000	61,841	40,010	61,831	73,584
	21–28	94,738	40,000	54,738	40,027	54,711	54,242
March	1–10	85,323	40,000	45,323	40,001	45,322	46,686
	11–20	75,967	40,000	35,967	38,644	37,323	40,192
	21–31	71,570	36,570	35,000	35,013	36,557	38,685
April	1–10	78,588	40,000	38,588	38,114	40,474	43,960
	11–20	90,955	40,000	50,955	35,003	55,952	53,241
	21–30	87,901	40,000	47,901	40,025	47,876	53,627
May	1–10	102,203	40,000	62,203	35,018	72,185	74,886
	11–20	122,062	40,000	82,062	40,000	82,062	92,039
	21–31	121,220	40,000	81,220	40,002	81,218	84,965

The river provides drinking water, sustains agriculture, forestry, fisheries, and inland navigation, helps to operate a quarter of the country's industrial activities, prevents salinity intrusion from the Bay of Bengal, and plays a determining role in maintaining the ecological balance of the country – especially that of the south-western region, which covers about a quarter of Bangladesh. The Gorai, which is the main tributary carrying water to the south-west region, now becomes totally dry at the beginning of the lean period.

Impacts of reduction of the Ganges flow in various sectors are interdependent and interlinked. In the following paragraphs sectoral overall damages are summarized. Some of these are quantifiable while others can only be discussed qualitatively.

Impact of reduction in supply of surface water

The economic life of the south-west region of Bangladesh is dependent on the Ganges. Of its distributaries, the Gorai plays the dominant role as it passes towards the industrial belt of Khulna, linking the Rupsa-Passur and Sibsa river systems and eventually emptying into the Bay of Bengal. The sweet water supply through the Gorai is vital for pushing back salinity and keeping an overall environmental balance. This distributary is also the source of potential irrigation development. The reduction of dry-season (January–May) natural flows in the Ganges in Bangladesh reduces the hydraulic efficiency of the channel to such an extent that even during high flows in the monsoon the progressive degradation of the channel and its hydraulic characteristics remains unabated.

Impact on groundwater

It has been observed that lowered river levels result in a reversal of the existing groundwater gradient, affecting the availability of groundwater. Since 1976 a fall in groundwater of about three metres has been observed in most of the wells along both banks of the Ganges, the Mohananda, and the Gorai-Modhumati from the pre-diversion normal level. The fall in groundwater level is greatest in the districts of Rajshahi and Pabna, followed by Kushtia and Jessore. The quality of groundwater has also deteriorated (Nishat 1996).

Channel morphology

The channel morphology of the Ganges and its distributaries has also been affected. A study of the longitudinal bed profile of the Ganges for the years 1974 and 1989 from its confluence with the Brahmaputra to the Indo-Bangladesh border and a study of changes at several cross-sections between the years 1974 and 1991 revealed the fact that the bed of the Ganges has silted up substantially in the recent past.

Impact on navigation

A total of 685 km of waterways which were navigable during pre-diversion time have been adversely affected during the post-diversion period. The confluence of the Ganges and the Brahmaputra was affected by shoals. As a consequence, the BIWTA (Bangladesh Inland Water Transport Authority) ferries are facing severe problems and dredging requirements have increased.

Increases in salinity

The most devastating effect of the diversion of the Ganges water has been caused by the marked increase in salinity in both surface water and groundwater, leading to higher soil salinity in the south-west region of Bangladesh. The major direct adverse effect of salinity is felt in industry, agricultural production, fisheries, forestry, and power generation. In addition, and most significant, such marked increases in salinity have short- and long-term impacts on health, expected mortality, and the ecosystem as a whole. Since the Farakka withdrawals commenced, the salinity ingress pattern in the area has a tendency towards cumulative increase due to residual deposits; this would be further aggravated if the present pattern of Ganges flow continues.

Reduction in dry-season Ganges flow raised the salinity of the Khulna area from 380 micro-mhos/cm during the pre-diversion period (1974) to about 29,500 micro-mhos/cm in April 1992. The salinity front of 500 micro-mhos/cm moved through the Passur estuary from 145 km to about 219 km inland after the diversion. This has resulted in increased soil salinity, leading to crop damage and severe yield reduction.

Impact on fisheries

The reduction of wetlands has resulted in reduction in fish catches. The most adversely affected fish are those which used to migrate upstream for breeding and spawning. Hilsa, which is an anadromous fish that requires freshwater flow, used to migrate upstream to spawn. As a result of the barrage, this fish is no longer available in India and is in decline in Bangladesh.

Impact on domestic and municipal water supply and public health

The Ganges river is the main source of domestic and municipal water supplies to the Ganges-dependent area in Bangladesh. The people use both surface water and groundwater for domestic purposes. Due to the drastic reduction of surface water, the people have been totally dependent on groundwater. But the availability and quality of the groundwater have become constrained too, due to lowering of the groundwater table and salinity intrusion. During the pre-diversion period, the water quality was well within the acceptable limits; later these limits were regularly violated. Increased salinity in the area has caused adverse effects on the general health of the people living there. The effect of salinity has given rise to an increased incidence of various ailments among the people using waters in the saline-affected areas. Waterborne diseases like typhoid, infectious hepatitis, diarrhoea, cholera, etc. among the inhabitants of the south-western region have been reported in increasing numbers.

Impact on forestry

The Sundarbans, littoral mangrove forests adjacent to the Bay of Bengal and comprising an area of 1,006,000 acres in Bangladesh, are situated in the south-west of Khulna district. The forests extend about 80 kilometres north of the Bay of Bengal, and are bounded on the east by the Baleswar river and on the west by the international boundary with India. With the increase of salinity, sundari, the main species of the Sundarbans, started dying and the regeneration of the species also decreased. Sundari may ultimately disappear if the salinity goes above the tolerance limit.

Impact on agriculture

Shortage of water in the Ganges since the commissioning of the Farakka barrage in India has adversely impacted crop production in a substantial area where soil moisture depletion has led to water stress. Expansion of irrigation facilities in the area served by the Ganges suffered heavy setbacks that retarded growth in agriculture. The damages in the agriculture sector due to reduced dry-season flows during the entire post-diversion period are manifold, and have been quantified for individual causes like shortage of water in the crop fields, soil moisture depletion, soil and water salinity exceeding tolerable limits, a lowered groundwater table, and delayed planting.

Impact on ecology and environment

The flora and fauna of south-western Bangladesh have evolved in response to natural conditions to form a climax ecosystem, which is now threatened with degradation caused by the inadequate supply of fresh water. The actions which triggered the degradation process of the total ecosystem in this region can be grouped into two broad categories: salinity increases in soil and water (surface and ground) from tidal ingress; and accelerated siltation in channels. In specific terms, the manifestations of reduced water flow are in the form of the northward penetration of tidal water causing salinity increases in soil and water; the diminished potential of the Sundarbans mangrove ecosystem; massive siltation and shoaling resulting in drainage congestion; decreased navigational draft in inland waterways; degradation of wetlands and reduction of open-water-capture fisheries area; and deterioration of groundwater quantity and quality.

Steps towards offsetting these adverse effects

Increasing fresh water supply in the affected area is considered to be the main technical solution to these problems. Now that a definite quantum has been secured for the lower riparian, steps will soon be taken to restore flows through the river network of the area. Towards this the following phases are envisaged:

- removal of sediment deposit from the month of the Gorai;
- construction of a control structure on the Ganges at a suitable location;
- restoration of the network of rivers, the conditions of which have deteriorated significantly.

Potential for regional approaches to water resources development

The Ganges and the Brahmaputra have basins spread over China, Nepal, India, Bhutan, and Bangladesh. They have distinctly separate basins, but they drain out to the Bay of Bengal via a common outlet, known as the Meghna estuary. The waters of the Brahmaputra have not been harnessed so far, but the Ganges basin is highly developed, with a large number of dams and barrages constructed in India and Nepal.

Besides these two major rivers there are about 54 common rivers between Bangladesh and India. The Meghna (known as the Barak in India) is also considered to be a major river. Among other rivers the Dharla, the Dudhkumar, the Teesta, the Monu, the Muhuri, the Khowai, the Gumati, the Mohananda, the Ichamati, and the Someshwari may be termed as medium-sized rivers while the rest are minor or small channels. All these watercourses may be termed as international rivers as they flow through two countries. Utilization of water resources of these rivers, and construction of bank protection and river training works in the border region, have been causes of dispute between the two countries. The main attention has, however, been on the Ganges.

Both the Ganges and the Brahmaputra have sufficient discharge on an annual basis to meet the present as well as the future water demands of the entire region. But in the dry season the flows become too small for meeting even the present requirements. Thus at one stage India and Bangladesh agreed to explore methods of augmentation of the dry-season flows of the Ganges. Bangladesh proposed conservation of the huge monsoon flows through storage reservoirs in the upper catchments in India and Nepal. India suggested diversion of 100,000 cusecs of flow of the Brahmaputra into the Ganges upstream of the Farakka barrage through a 200 km link canal. The proposals of one country have not been acceptable to the other for various reasons.

If an agreement could be reached on how to harness and utilize the water resources of all the rivers of the region effectively, tremendous benefits could be obtained in terms of irrigation facilities, hydropower generation, navigation routes, flood mitigation, fish production, and management of the environment and ecosystem. There is a shortage of water in the dry months, but the huge monsoon flow remains untapped. If a beginning could be made then the rivers of the region could become waters of hope, as aptly termed by Verghese (1990). Water is there, but not when needed.

A major dimension of the management and utilization of the GBM waters pivots around the concept of storage reservoirs in the upper reaches of the Ganges. The construction of reservoirs in upstream reaches for optimal and multiple use of water resources of the region could achieve flood-peak attenuation in lower reaches, augment dry-season flows, generate hydro-electricity, and offer irrigation as well as navigation opportunities. There are a number of good sites for creating reservoirs in the Ganges tributaries in the middle mountain belt of Nepal. Studies carried out by Bangladesh in 1978, and updated in 1983, outline a scheme for optimum development of the surface water resource of the Ganges basin through a coordinated and cooperative plan implemented by the co-basin countries – Nepal, India, and Bangladesh. The studies suggested construction of seven storage reservoirs in Nepal as well multiple storage dams in India, with a view to holding vast monsoon runoff and increasing dry-season availability of the Ganges waters. Needless to say, these mega-constructions need huge investments of capital, and therefore the objective would have to include multiple uses accrued from the storage reservoirs which could be shared cooperatively by the co-basin countries.

Joint efforts in flood management

Floods are a recurrent problem in the GBM region, especially in Bangladesh and eastern and north-eastern India. Every year substantial damage is caused by floods in Nepal, India, and Bangladesh. Nepal's worst flood in six decades occurred in 1993, when over 2,000 people lost their lives. India also experienced one of its worst floods of the century in 1993 in the northern states. In 1988, nearly 60 per cent of Bangladesh's land area was submerged in widespread flooding of catastrophic proportions. Any mitigation plan for flood hazards needs mutual cooperation and efforts.

Flood forecasting

Flood forecasting is an important non-structural flood mitigation measure with much scope for inter-country cooperation. It can not only safeguard life and property in unprotected areas, but also prevent damage to engineering structures. In Bangladesh, the Flood Forecasting and Warning Centre of the Bangladesh Water Development Board (BWDB) is responsible for issuing flood warnings. The centre has a network of wireless stations for monitoring real-time water levels and rainfall data. It also has links with the Bangladesh Meteorological Department, and, during flood time, it receives satellite imagery from the Space Research and Remote Sensing Organization (SPARRSO). Under a joint-action programme of flood forecasting and warning, Bangladesh and India have had an agreement for transmission of real-time and forecast flood data since 1972. The present arrangement of transmission of data (whenever the water level is at or above warning stage) from India is limited to such stations as the Ganges at Farakka, the Brahmaputra at Goalpara and Dhubri, the Teesta at Domohoni, and the Barak (Meghna) at Silchar.

Current arrangements for flood forecasting data transmission do not allow lead times of more than 24 hours for central Bangladesh – and even less for the border areas. The cooperative arrangement between Bangladesh and India may be strengthened through making provisions for three-hourly real-time and daily forecast data transmission during the monsoon months of May to October, irrespective of warning stage. In order to increase the forecast lead time in Bangladesh, the data should be transmitted from more upstream stations on the Ganges, Brahmaputra, Barak (Meghna), and Teesta. Arrangements can also be made between Bangladesh and Nepal for transmission of real-time water-level, discharge, and rainfall data which would be very useful for early warning of floods in the Ganges and disaster preparedness measures in downstream sites. Similar arrangements between Bangladesh and Bhutan would be useful for early warning of floods in the Brahmaputra.

Embankments

Embankments are an important structural intervention to achieve the objective of flood protection, and inter-country cooperation in this sector can be envisaged for the border rivers. Embankments have been constructed in Bangladesh for many years for protection against flood, and the ones which have been properly planned, constructed, and maintained have functioned well. In order to make them more effective, the embankments on both banks of all the common rivers must be tied up with those beyond the borders. Some embankments are already tied, but through mutual agreement and collaboration the gaps in the embankments along the common rivers like the Ganges, Brahmaputra, Dharla, Dudkhumar, Khowai, and others could be identified, and tying up could be completed in phases.

River training

River training is another engineering effort to prevent bank erosion, ensure flow regulation, protect settlements and infrastructures, maintain navigability in channels, and control sediment flow. The GBM countries may exchange experience and expertise in collaborative study and research in order to evaluate the various methods of river training in their respective countries.

Apportionment of waters of common rivers

Apportionment of waters from the rivers in the GBM basins (except for the Ganges between India and Bangladesh and the Mahakali between India and Nepal) is still an unresolved issue. With regard to the left bank tributaries of the Ganges basin, Nepal is the upper riparian in all except for a small stretch of the Mahakali, where it is a common river. On the other hand, India is the upper riparian of all the three river systems of Bangladesh – the Ganges, Brahmaputra, and Meghna. Being the lower riparian, Bangladesh offers an outlet for the entire volume of the three rivers in the flood season, but remains dependent on the upper riparian for the lean-season flows. Hence, a cooperative and equitable agreement to share fresh water resources in the GBM region is essential.

Hydropower generation and trade

All the four countries of the GBM region, namely, Bangladesh, India, Nepal, and Bhutan, have a per capita consumption of commercial energy much lower than the global average. The most important characteristic of the energy economy of these countries is their high dependence on non-commercial energy – mainly derived from biomass. Non-commercial energy is extensively used in rural areas and by the urban poor, while commercial energy is largely urban-centred – confined to urban households, transportation, and industrial sectors.

The abundant supply of water and the favourable terrain offer excellent opportunities for hydropower development in Nepal, which can meet the energy requirements for the GBM region for many years to come. Hydro-electricity is environmentally friendly, clean, and renewable – hence, cooperative efforts in hydropower generation and its trade in the GBM region are another potentially important sector of regional cooperation. The installed capacity of electricity production in Nepal is about 250 MW, of which 227 MW is generated from water and the rest from diesel. The volume of water in Nepal's rivers has a theoretical hydro potential of 83,000 MW, of which the economically feasible energy potential is about 42,000 MW. The current production capacity (227 MW) is only 0.27 per cent of the theoretical potential – which underscores the enormous hydropower potential of Nepal.

Nepal's development of hydropower utilization had been slow due to the small size of the country, financial constraints, and the limited nature of the domestic market. However, the domestic demand for power is increasing, and currently Nepal does face a power shortage. The huge hydropower potential of the upper Ganges in Nepal must be assessed in the context of immediate geographic realities in the region. Besides the rising domestic demand, there is also an increasingly hungry market for power across the southern borders of Nepal. The ideal combination of an energy potential at home and a market for energy export in the neighbouring region is a sufficient justification for the exploitation of Nepal's water assets. The logical market is northern India, especially the states of Bihar and Uttar Pradesh. It is also feasible that, in the long run, Nepal might be able to transmit electricity across India into Pakistan and Bangladesh. However, the immediate potential for Nepal's hydropower exists in the northern region of India, where – instead of constructing polluting thermal plants to meet rising energy demand – India can buy hydropower from Nepal. An ultimate cooperative arrangement could be envisaged through the interconnection of the grids of Nepal and northern India.

Navigation and access to the sea

Fuller utilization of water resources in the GBM region should also include an examination of the opportunities for cooperation in inland navigation and river transit. Nepal and Bhutan are two landlocked countries of the region, and are naturally interested in transit through another country for access to the sea. First, storage reservoirs in the upper Ganges could offer opportunities for improving the draft for inland water transport. Second, transit facilities for Nepal through the Ganges waterway or a link with the Jamuna-Padma system could be explored to give the country access to the Bay of Bengal.

Nepal has transit treaties with both Bangladesh and India. Most of Nepal's overseas trade is channelled through Calcutta, and greater cooperation in this field would undoubtedly provide an impetus to Nepal's economy. Feasibility studies should be carried out to explore the possibility of utilizing more upstream locations in the Ganges, which are closer to the Nepalese border and might be suitable for internodal cargo transfer.

The inland waterways of Bangladesh provide a theoretical opportunity for water transit facilities to West Bengal and north-eastern states of India. The potential for utilizing the navigation routes for transit through the Brahmaputra, Padma, and Meghna may be examined. However, before this opportunity can be realized, a complex host of politico-economic and security problems have to be solved. In a similar vein, one can also talk of linking the rivers of Bangladesh with those of eastern Nepal, which could provide significant transit potential for Nepal's commercial traffic to the outside world. An early proposal indicated the feasibility of constructing a 470 km water route linking Biratnagar in eastern Nepal with Bera in Pabna (Bangladesh), whence the traffic would utilize the existing waterways through the Padma and Meghna, and reach Mangla (south-western Bangladesh) via Chandpur. This water route requires the construction of canals to link with existing waterways in north-western Bangladesh, as well as some essential river training works and rail and road crossings. Nonetheless, the route remains an alternate maritime exit for Nepal, and it could also boost trading contacts between Nepal and Bangladesh.

A technical committee on transport under SAARC (South Asia Association for Regional Cooperation) sponsorship is understood to have deliberated on the improvement of transport infrastructure and transit facilities among the neighbouring countries. In the context of the changed and liberal global economic environment, it is perhaps not too much to expect that the GBM countries would also demonstrate a willingness to cooperate in the sector of inland navigation to their mutual benefit.

Catchment management

Catchment is a hydrologic unit of land area, and its dimension is a direct function of the aerial extent of the drainage basin/basins. Over 90 per cent of the catchments of the three major river systems of Bangladesh fall outside its territory; hence, Bangladesh has a natural concern for an objective and planned management of these catchments. Catchment management in the GBM region essentially involves forest management on a macro scale in order to rehabilitate badly denuded slopes, reduce runoff, and prevent soil erosion. The progressive rate of deforestation in the Himalayan slopes has exacerbated the problem of sedimentation in the river channels, which has been one of the contributory factors to increased flooding in downstream areas. Hence, it is beyond dispute that sound catchment management in the GBM region is imperative through programmes of soil conservation, reforestation, catchment area treatment, and strict land-use control. Bangladesh would be interested to share Indian, Nepalese, and Pakistani experiences in catchment management. Joint cooperative study, research, and investigation programmes could be undertaken among the SAARC countries for appropriate catchment management with a view to mitigating the adversities of topsoil erosion, landslides, sedimentation, and unplanned farmland encroachment of slopes.

Disaster management

The principal strategy of disaster management is disaster preparedness, with the objective of mitigating losses. Bangladesh has a comprehensive programme of disaster management aimed at dealing with such natural hazards as floods and tropical cyclones. A Disaster Management Bureau was formed in the Ministry of Disaster Management and Relief in 1993 with the prime objective of making people in disaster-prone areas aware of the hazards so that loss of life and property can be minimized. The bureau's primary concern is with the pre-disaster period, and it is also in the process of reviewing and updating the national guidelines – known as standing orders – for disaster management. In order to provide national coverage for its activities and programmes, the bureau has already formed a hierarchy of disaster management committees from the national level down to district and union levels. Similar disaster management bodies also exist in various levels of operation in other SAARC countries. These might be concerned not only with floods and cyclones, but also other natural hazards like earthquakes, glacial lake outbursts, snowfall, avalanches, tornadoes, etc. All the SAARC countries would benefit immensely from the experience of each other in facing disasters through an exchange of information and expertise in their tasks of protecting the population from natural hazards.

Concluding remarks

The management of fresh water which crosses international borders is a matter of inter-country concern – it is a transboundary environmental resource and the environment does not recognize borders. The regional cooperation in water management in the sectors outlined above is, therefore, a quintessence for the GBM countries' sustainable development. This region is characterized by mass poverty and economic stagnation. Cooperative endeavours in the water sector could be a starting instrument for South Asia in the path toward self-reliance and sustainability. The bases of cooperation already exist. A multilateral body for water development in the GBM region may be formed under the SAARC umbrella with a mandate to examine and recommend measures of cooperation on the basis of social justice and equity. With positive thinking and respect for each other's rights and legitimate needs, it should not be difficult to identify avenues and modalities for cooperative ventures. There are several examples of cooperative water utilization programmes in transboundary river basins around the world which can be gainfully replicated in the GBM region. All that are needed to begin the task are a genuine desire to attain that goal and an agreement to reach a solution where everyone will be a winner and nobody will be a loser. The time has come to think about the institutional framework. Should we continue bilaterally or move towards a regional institution? This would require a significant adjustment in the political framework of the countries of the region.

REFERENCES

Abbas, B. M. 1982. *Ganges Water Dispute*. Dhaka: University Press.

Begum, Khurshida. 1987. *Tension Over the Farakka Barrage*. Dhaka: University Press.

Crow, Ben. 1995. *Sharing the Ganges: The Politics and Technology of River Development*. New Delhi: Sage Publications.

Nishat, Ainun. 1996. "Impact of Ganges Water Dispute on Bangladesh", in *Asian International Waters, From Ganges-Brahmaputra to Mekong*, eds Asit K. Biswas and T. Hashimoto. New Delhi: Oxford University Press.

Verghese, B. G. 1990. *Waters of Hope*. New Delhi: Oxford/IBH Publishing.

Verghese, B. G. and Ramaswamy R. Iyer. 1993. *Harnessing the Eastern Himalayan Rivers*. New Delhi: Konarak Publishers.

6

Harnessing the Himalayan waters of Nepal: A case for partnership for the Ganges basin

Iswer R. Onta

Introduction

General

Balanced development of water resources requires a multidisciplinary, multisectoral, integrated river basin approach. Since the water resources are shared by several groups and stakeholders within each river basin, the implications of each action affecting water quality and quantity must be taken into account. In the past, the perceived water development needs of various river basin sectors and stakeholders have been addressed separately by different governments and agencies.

The Ganges is an international river originating in the Himalayan peaks and draining off ultimately to the Bay of Bengal, with its basin spread over four sovereign nations: China, Nepal, India, and Bangladesh (Figure 6.1). The case of Tibet will not be discussed here as the potential for using the water resources is extremely limited there, be it in irrigation, hydropower, or any other sector.

The Ganges basin has a total area of 1,087,300 square kilometres. The proportions of basin area in China, Nepal, India, and Bangladesh are 3.08, 13.56, 79.10 and 4.26 per cent, respectively (India and Bangladesh also share the Brahmaputra basin, which has a total area of 552,000 square kilometres, in the proportions of 35.33 and 7.08 per cent respec-

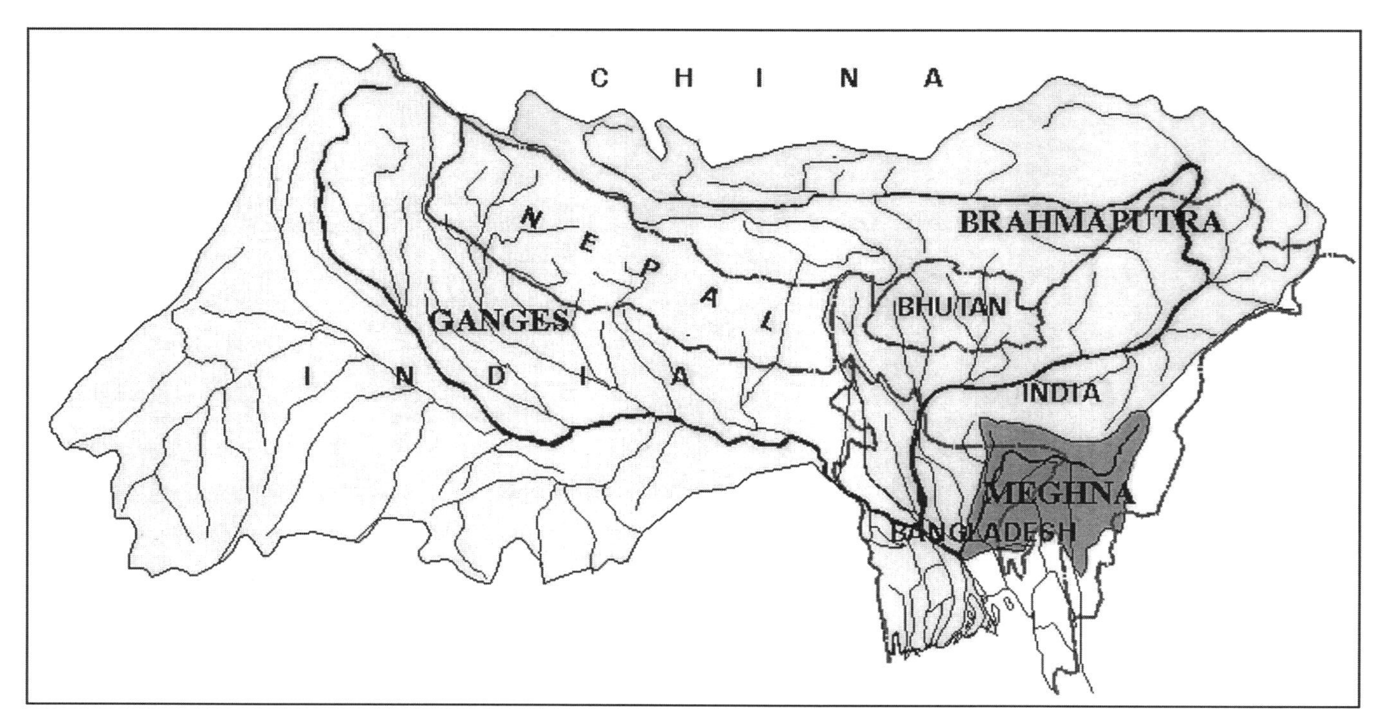

Figure 6.1 The regional setting – the Ganges, Brahmaputra, and Meghna river basins

Table 6.1 Ganges basin area distribution

Country	Basin area (km^2)	% of total area
China	33,520	3.08
Nepal	147,480	13.56
India	860,000	79.10
Bangladesh	46,300	4.26
Total	1,087,300	100.00

Source: Shrestha and Singh 1996

tively). The distribution of the Ganges basin area in each country is shown in Table 6.1.

The population of the Ganges basin is about 344 million people (270 million in India, 52 million in Bangladesh, and 22 million in Nepal). However, the Ganges-Brahmaputra basin has over 535 million inhabitants.

Water in the Ganges basin is abundant, and if properly developed and managed the region could really come out of the poverty trap it is in and start on a growth track. The so-called intractable issues hindering the optimal development of water resources in the Ganges basin over the last 50 years or so have been a subject of intense intellectual and technical discussions. The lesson learnt so far is that the best way to look at any given river system is from the source to the sea: the entire river basin should be viewed with a holistic and sustainable approach rather than a sectoral one (Biswas and Hashimoto 1996).

The situation becomes more complex when the basin is shared by more than one sovereign state (as is the case here), with different socio-economic and political realities and national interests and priorities (Bhadra 1997). The complexity is further increased when a co-riparian country has other major river basins within its territory, with the possibilities of inter-basin water transfers. This is the case for India and Bangladesh in the sharing of waters of the Ganges and Brahmaputra basins (Char 1997).

After the liberalization of economic measures taken a few years back in India, coupled with the democratization of polity of Nepal and Bangladesh in early 1990, serious dialogues among water professionals, bureaucrats, politicians, and the press have been initiated; this is a continuing process. Encouraging debates and discussions have taken place, and there are signs that a consensus has been reached on harnessing the potential of the Ganges and its tributaries for the overall balanced development of the people of the basin (Prasad 1997).

Overview of the Nepalese socio-economic scenario

Nepal is a landlocked country in Asia, bordered by China to the north and India to the south, east, and west; it is located between latitudes 26° 22′ N–30° 27′ N and longitudes 80° 4′ E–88° 12′ E. The country, covering a total area of 147,480 square kilometres, has very diverse physical characteristics in its topography, climate, geology, and land-use system. It is predominantly mountainous, with three broad agro-ecological zones: the terai plains (including the inner terai), hills, and mountains (including the Himalayas), covering respectively about 18, 58 and 24 per cent of the total land area. Administratively, Nepal is divided into five development regions, 14 zones, and 75 districts.

Nepal is one of the poorest and least developed countries of the world, with an ever-growing population of about 22 million (estimate of mid-1996), and an annual growth rate of 2.5 per cent. Only about 8 per cent of the population live in the mountains, the rest being almost equally divided between the hills and the terai. The annual per capita gross national product (GNP) is only US$220, and the corresponding growth rate in per capita output has remained very low (less than 2 per cent) for many years.

Agriculture is the mainstay of the economy, contributing about 42 per cent to the gross domestic product (GDP) according to recent available estimates, and providing employment to more than 80 per cent of the working population. However, the performance of this sector has been less than satisfactory, with productivity declining or remaining stagnant over the years. The industrial sector contributes a little more than 22 per cent to the total GDP, and the services sectors contribute 35 per cent to the GDP. The data reveal that the structural transformation of the Nepalese economy has been minimal.

According to recent estimates, about 42 per cent of the Nepalese population have income levels below the poverty line. The major reason for such poverty is the lack of productive assets in the form of arable land. The average size of land-holding per cultivator household is less than 1 ha in the terai region and 0.3 ha in the hill region. Per capita arable land declined to 0.12 ha in 1996 from 0.60 ha in 1964.

The urban population has grown to 14 per cent, compared to 26 per cent in South Asia overall. Life expectancy at birth in Nepal is 55 years, whereas it is 61 years in South Asia. The infant mortality rate per 1,000 live births is 91 in Nepal, compared to 75 in South Asia. Access to safe water is available to only 60 per cent of the population and illiteracy is rampant, with 73 per cent of the population above the age of 15 being illiterate.

The total GDP of Nepal is estimated at US$4.5 billion; the average annual growth in GDP is estimated at 4.7 per cent. Despite a modest increase in the real GDP, per capita income has remained more or less stagnant over the last two decades due to the high population growth.

At present, it is estimated that the total cultivated and irrigable lands are 2.64 and 2.18 million ha respectively. However, only 1.77 million ha represent suitable agriculture areas with irrigation potential. The majority, 1.34 million ha (76 per cent), lies in the terai, while the remaining suitable irrigable area is mainly located in the hill districts (Table 6.2). It is also estimated that, by the end of 1992, irrigation infrastructure had been developed by different agencies for about 1.09 million ha. It is estimated that 0.77 million ha (71 per cent) and 0.26 million ha (24 per cent) of these facilities exist in the terai and the hills, respectively. Irrigation is mainly supplementary during the monsoons, and year-round irrigation is available in much less than 50 per cent of the irrigated 1.09 million ha.

Water resources of Nepal

By and large Nepal's water resources are considered to be abundant. The average annual precipitation is about 1,500 mm (80 per cent of which occurs during the monsoon season from June to September). Although the rainfall is highly variable in specific locations, it is normally less in the terai than in the hills, and higher in eastern Nepal than in western Nepal. The total annual average runoff from Nepal is over 200 bcm. More than two-thirds of the annual runoff occurs during the monsoon period. Surface water resources are distributed in the river system, which consists of four major rivers, seven medium rivers, and a large number of small rivers. The spatial and temporal distribution of surface water resources, particularly due to the lack of storage facilities, creates surpluses at certain times and places and shortages at others. Besides surface water, Nepal is also endowed with extensive groundwater resources, mainly in the terai. Both shallow and deep aquifers may be utilized independently or in conjunction with surface water for irrigation purposes (Hillel 1997).

The major rivers originating from the snow-capped Himalayas are the Mahakali, the Karnali, the Gandaki, and the Saptakosi. The second group of medium rivers originate in the middle mountains, and are mostly rainfed and have low dry-season flows. The rivers in the third category are small catchments originating from the Churia hill ranges of Nepal. These rivers will have very little or no flows at all during the dry season. The total number of rivers is estimated at 6,000, and the total length of the water courses is about 45,000 km.

The mean annual flows from the Himalayan rivers, including flow from

Table 6.2 Distribution of irrigated and potentially irrigable land

				Potential irrigable land (net ha)			
Ecological belt	Irrigated area (ha)	Rainfed area (ha)	Total cultivated area (ha)	Total irrigable agricultural area	Irrigable non-agricultural area	Total irrigable area	Total remaining irrigable area
Terai	780,900	578,100	1,359,000	1,338,000	406,000	1,744,000	963,100
Hills	199,400	855,600	1,055,000	368,000	5,000	373,000	173,600
Mountains	30,700	196,300	227,000	60,000	1,000	61,000	30,300
Total	1,011,000	1,630,000	2,641,000	1,766,000	412,000	2,178,000	1,167,000

Source: *Master Plan for Irrigation Development in Nepal*

Table 6.3 Estimated mean annual runoff – Nepal rivers

Category of basin	River	Basin area (km^2)	Estimated annual runoff (m^3/sec)
Himalayan	Mahakali	15,640	656.88
	Karnali	44,000	1,430.00
	Gandaki	34,960	1,779.46
	Koshi	60,400	1,673.08
Mahabharat	Babai	3,252	82.28
	West Rapti	6,215	148.54
	Bagmati	3,681	217.91
	Kamala	1,786	52.15
	Kankai	1,317	65.06
Churia	(Various)	22,797	456.00
Total			6,561.36

Source: Institute for Integrated Development Studies 1995

outside the Nepalese territory, are estimated at about 5,539 m^3/sec. The total runoff from the medium rivers of Nepal is estimated at 566 m^3/sec and the runoff from the Churia rivers is estimated at 456 m^3/sec (Table 6.3). Overall, the rivers of Nepal contribute more than 40 per cent of the total flow of the Ganges and over 70 per cent of its dry-season flow. Water storage potential in Nepal is 88 bcm. If properly planned and managed, development of this storage potential could yield tremendous benefits to Nepal and its neighbours in the form of hydropower generation, flood control during the monsoon, flow augmentation for downstream irrigation and navigation, and water supply. Hence it is natural that India has a vital concern about Nepalese waters and will have keen interest in Nepal's water resources development.

Overview of water resources development in Nepal

Water is needed for domestic consumption, food production, energy generation, industry, navigation, recreation, and sanitation.

Drinking water supply

Sixty per cent of the population of Nepal is assumed to have access to potable water, and it is estimated that at present 674 million litres of water per day are supplied to this population. A piped water supply system has been installed in all the 75 district headquarters of Nepal. The government of Nepal intended to provide drinking water facilities to 72

and 77 per cent of the rural and urban population, respectively, by the end of 1997.

The government had ambitious long-term plans to provide drinking water to the entire population by the turn of the century. However, the seasonality of flows has continued to impede the availability of safe and reliable water supplies, and past investments in infrastructure have not had the developmental and social impact originally expected.

Irrigation development

As mentioned earlier, out of 1.77 million ha of potential irrigable land, irrigation infrastructure has been developed for over 1.09 million ha. Large-scale irrigation is feasible only in the terai plains, where extensive and more uniform land is available. Most possible run-of-the-river projects have already been taken up for development, though intensive infrastructure development and quality water management are lacking.

Multipurpose projects like the Kankai, the Sunkosi-Kamala diversion project, the Bheri-Babai diversion project, the Karnali, etc., which have high potentials for year-round irrigation in the terai, are awaiting implementation (Onta, Agrawal, and Onta 1995).

Groundwater potential (the dynamic groundwater reserve is estimated at 8.8 bcm) for irrigation development in the terai is also very high, and the Agriculture Perspective Plan (APP) of Nepal envisages irrigation of 0.5 million ha of the terai land by exploiting this potential.

Farmer-managed irrigation systems (FMIS) have been operating successfully in the terai, hill, and mountain areas of Nepal for centuries. These systems have been rediscovered, and the government has adopted the FMIS community management system as a focus to manage irrigation development by incorporating it into its irrigation policy.

Hydropower situation

Nepal's theoretical hydropower potential is enormous, at 83,000 MW (identified power potential is 42,000 MW – see Table 6.4). However, to date, Nepal has developed just about 244 MW of its vast hydropower potential. The population having access to electricity was estimated at 12 per cent in 1996. At present, nearly all urban and semi-urban population centres have been electrified: 89 per cent of the urban population and only 4.5 per cent of the rural population receive electricity services.

Nepal has classified the hydropower schemes into three categories: small (up to 10 MW), medium (up to 500 MW), and large (more than 500 MW). The Nepal Electricity Authority (NEA) has classified the hydropower schemes into four categories: micro (up to 1 MW), small (1–10

Table 6.4 Identified hydropower potential in Nepal

River basin	No. of sites	Identified power potential (MW)
Sapta Koshi	40	10,860
Gandaki	12	5,270
Karnali	7	24,000
Mahakali	2	1,125
Other southern rivers	5	878
Total	66	42,133

Source: Institute for Integrated Development Studies 1995

MW), medium (10–300 MW), and large (more than 300 MW), to reflect the present situation and future requirements. The small-size schemes are designed to serve local grids in remote locations by harnessing numerous small rivers; the medium size to cater to national needs and export the excess; and large schemes mainly for export to neighbouring countries.

Currently, hydropower schemes with a total capacity of 280 MW, including the 144 MW Kaligandaki A project, are under construction. All of these schemes are run-of-the-river type and most should be available in the system by the year 2000. The next lot of hydro schemes, with generating capacity of 385 MW including the 240 MW Upper Karnali, are being studied at various levels and will meet the projected domestic electric energy needs, if implemented, till 2007–2010 depending upon the forecasts made by the NEA at different times in different situations (Table 6.5).

Other sectors

Development in other sectors related to water resources, including navigational needs, has been minimal and at the most potential projects and schemes have been identified.

Past experience in large-scale water resources development

The fact that the Himalayan rivers originate in Nepal holds a large potential for the development of hydropower as well as irrigation facilities, and to some extent inland water transport has been known to planners, scholars, and developers since the early twentieth century. The quantification of resources, which started to emerge only during the later stages, is currently being fine-tuned by governments of the region as well as by NGO research activities.

Table 6.5 NEA load forecasts

Year	Peak load (MW)	Energy (GWh/yr)	Load factor (%)
1996	281	1,272	51.7
1997	303	1,392	52.4
1998	328	1,530	53.2
1999	354	1,688	54.4
2000	383	1,860	55.4
2001	413	2,047	56.6
2002	452	2,261	57.1
2003	498	2,500	57.3
2004	549	2,767	57.5
2005	610	3,065	57.4
2006	676	3,392	57.3
2007	748	3,757	57.3
2008	824	4,153	57.5
2009	909	4,593	57.7
2010	1,002	5,081	57.9
2011	1,104	5,617	58.1
2012	1,215	6,216	58.4
2013	1,340	6,879	58.6
2014	1,483	7,641	58.8
2015	1,642	8,487	59.0

Source: Nepal Electricity Authority 1996

Past Indo-Nepal cooperation over international rivers

Nepal-India cooperation in large-scale water resources development dates back to the early years of this century (1920, during the British rule in India). The Sarada barrage on the border river Mahakali (known as the Sarada in India) was constructed by India after exchanging some land between Nepal and India. The barrage is situated wholly in India, and Nepal is entitled to a specified supply of water for irrigation.

After an interval of over 30 years, Nepal and India signed an agreement to construct the Kosi barrage in 1954. The agreement to construct the Gandak barrage was signed between Nepal and India in 1959. Both these barrages were constructed on or close to the Nepal-India border and were based on India's initiative and needs. They were wholly financed by India and mostly benefited India (with the only benefit to Nepal being a specified supply of water for irrigation and entitlement to power).

These early Indo-Nepal water resources projects, though considered reasonable from India's viewpoint, were seen as a "sell-out" by many in Nepal. Other water resources development in rivers in India close to

Nepal's border has been perceived as "not-so-friendly activities" by Nepal. The consequence has been such that the constitution of Nepal embodied Article 126, requiring any "treaty" pertaining to natural resources and certain other matters to be ratified by a two-thirds majority by the country's parliament, to safeguard Nepal's national interests. However, with the introduction of a liberal outlook in the economy in India in the early 1990s, and with the restoration of multi-party democracy in Nepal as well as in Bangladesh, the climate of economic co-operation and water resources exploitation has seen marked changes recently.

Issues related to regional cooperation

India needs the hydropower resources of Nepal: by developing its major rivers, Nepal can provide large hydro and storage projects to augment the low flows during the lean season as well as to moderate the flood situation. Northern India remains short of power to the tune of more than 50,000 MW, but Nepal can help mitigate this situation to a considerable extent (Institute for Integrated Development Studies 1997).

Bangladesh wants Nepal to construct large storage dams to regulate the lean-season flow of the Ganges and augment the Ganges water so that the needs of both India and Bangladesh in lean seasons could be catered for. Nepal, in return, would enjoy considerable benefits from selling hydropower to India and Bangladesh, and would also be compensated for the use of its reservoir dam sites for augmentation of low flow as well as flood control. Nepal would also like to benefit from enhanced/developed inland waterways in one of its major rivers, mainly the Kosi, to have access to the sea for its export trade. The issue is how best all these needs can be catered for and how best to achieve a "win-win" situation for all the parties involved. This is the problem that needs to be solved.

Potential schemes to address the immediate issues

Though it would have been a wise step to develop separate strategic master plans for each of the important river basins, such as the Mahakali, the Karnali, and the Kosi, for their long-term optimal utilization, identified projects like the Pancheswar high dam on the Mahakali, the Chisapani high dam on the Karnali, and the Sunkosi high dam on the Saptakosi cannot be ignored. The siting of these high-dam projects could be optimal even if the whole of each basin is considered (though rigorous study of the basins could prove otherwise).

Table 6.6 Pancheswar project specifications

Dam height	315 m
Installed capacity	6,480 MW
Irrigation benefits to Nepal	93,000 ha
Cost	US$2.98 billion (1995 prices)
Live storage capacity	6.5 bcm

The Mahakali Pancheswar project

The Pancheswar high-dam project is part of the Mahakali river integrated development treaty, which underlines the provision made under the Sarada agreement of 1920 to provide upgraded benefits to Nepal from the Tanakpur project (8 km upstream from the Sarada barrage), and stipulates provisions in connection with the development of any project in the Mahakali river. A specific project agreement will, however, be necessary as and when new projects, including the Pancheswar, are developed.

The Pancheswar project was first identified by the Central Water and Power Commission of India in 1956. A project report was prepared by India in 1971, proposing the construction of a 232-metre-high concrete gravity dam having an installed capacity of 1,000 MW (500 MW on each side of the river).

Nepal and India decided to proceed with field investigations at the feasibility level, with each country working within its own territory. Based on the findings of the field investigation work and the project definition study, Nepal prepared a detailed project report in September 1993. There appears to be some divergence in the concepts of Nepal and India regarding the parameters of the project, specifically the height of the dam and the installed capacity of the power plant etc. The parameters as developed by Nepal are given in Table 6.6.

The Mahakali treaty was concluded in January 1996 between Nepal and India. Nepal sent the draft DPR (detailed project report) to India on 3 July 1996. The Nepalese parliament ratified the Mahakali treaty on 20 September 1996. In May 1997, India raised the issue of the lower Sarada project (the Sarada Sahayak canal), situated 160 km downstream of the Sarada barrage. The Mahakali treaty became effective on 5 June 1997 after an exchange of treaty documents during the visit of the Indian prime minister to Nepal.

Several joint meetings between Nepal and India have been held, but these meetings could not agree on the joint DPR mainly because basic issues like equal rights on water and the Sarada Sahayak canal could not be resolved. The parliament stricture to agree on the joint DPR within six months of ratifying the treaty could therefore not be achieved. A great

Table 6.7 Karnali project specifications

Height of rock-fill dam	270 m
Installed capacity	10,800 MW
Irrigation potential in Nepal	191,000 ha
Cost	US$4.89 billion (1989 prices)
Live storage capacity	16.2 bcm

deal of effort would be necessary to proceed in line with the provisions made under the Mahakali treaty.

Karnali project (Chisapani dam)

This is one of the largest multipurpose projects, and has been under discussion between Nepal and India since the late 1960s. Various studies have been carried out and the latest detailed feasibility study, undertaken by a consortium of Canadian consultants, envisages the project with the parameters shown in Table 6.7.

In this case, India has expressed a preference for a lower dam height at Chisapani with a capacity of 7,000 MW.

The private developer ENRON has shown interest in developing the Chisapani dam project and has requested the government to grant permission to start necessary surveys. Recently, after several discussions and presentations made by ENRON, permission was withheld pending more inquiries.

Kosi high-dam project

Kosi is another major multipurpose project with flood control, irrigation, power, and navigation benefits; it has been under discussion between Nepal and India for a long time.

India proposed the project, to be sited near the Barahchhetra gorge, as far back as 1950. India prepared a feasibility report in 1981. Nepal identified the project during the preparation of the Kosi basin master plan in 1985.

The project proposed by Nepal has parameters which differ from the Indian proposal; both specifications are given in Table 6.8.

The height of the Kosi high dam in the Nepal proposal is lower in order to accommodate the Sunkosi-Kamala diversion project upstream near Kurule, which has the potential to generate a significant amount of power and irrigate substantial land in Nepal.

The latest expert meeting on the Kosi high-dam multipurpose project between Nepal and India was held in January 1997, when it was decided to conduct a detailed study of the project starting in mid-July 1997. The

Table 6.8 Kosi project specifications

	Nepal's proposal	India's proposal
Dam height	239 m	269 m
Installed capacity	–	3,300 MW
Gross storage capacity	8.5 bcm	–
Irrigable area	68,850 ha (in Nepal)	1,522,000 ha (in India)
Cost	US$2.7 billion (1985 prices)	US$4.6 billion (1991 prices)

study was to cover all aspects of benefits including irrigation, flood control, hydropower, and navigation, and the project was to be evaluated according to the principles established in the Mahakali treaty.

Regulatory provisions

The legal and regulatory requirements to harness and develop the water resources of Nepal have been given due consideration after the restoration of democracy in 1990. Without these instruments in place, it would have been mere rhetoric. The following instruments (enacted and incorporated) are indications of the Nepalese intention to face head-on the issues related to management and harnessing of the water resources of Nepal: the Water Resources Act (1992), the Electricity Act (1992), the Electricity Tariff Fixation Regulation (1993), the Water Resources Regulation (1993), the Electricity Regulation (1993), the Industrial Enterprise Act (1992), the Foreign Investment and Technology Transfer Act (1992), the irrigation policy (1992), and the hydropower development policy (1992). Nepal is planning to amend the Water Resources Act, the Electricity Act, and the hydropower development policy to suit the current investment climate as well as other challenges of the future, including large-scale export of electrical power and downstream benefits arising out of storage projects.

Nepal's viewpoint

Nepal is of the view that its domestic energy needs will be met by utilizing small and medium hydropower schemes identified for a foreseeable future. Nepal has projects already under construction that will take care of its domestic need till 2005. Nepal has also identified schemes and studies are being conducted on other schemes to cater for its power needs for a few years beyond 2005. Some supply deficiency might occur during wintertime if peaking schemes are not undertaken to meet the peak demand.

Nepal has embarked upon the West Seti hydro-electric project (750 MW) with an agreement signed with a private developer, the Snowy Mountain Engineering Corporation (SMEC) of Australia, with the sole intention of exporting the substantial power to the Indian market. This is a storage project which could be used as a peaking station (if Nepal needs), and will be a test case in relation to power sales to India; it will also prove the efficacy of the power trade agreement that has been concluded with India.

High-dam projects, with the inherent risks associated with silt deposition, seismic activities, and environmental considerations, are not in the priority list of Nepal. These high-dam projects are also very expensive and investment funds are not likely to be acquired by Nepal. Besides, heavy investment in such schemes (which amount to more than Nepal's annual GDP) may not be a sound investment proposal (when competing projects in other sectors are available).

Though Nepal would like to earn revenue from selling power to India or Bangladesh to invest in important social sectors like drinking water, education, and health for long-term development, it will remain hesitant to make huge investments. At the same time, Nepal wants to extract its reasonable share from those proposed high-dam projects. Aspects like irrigation and flood control benefits are issues that cannot be left unaddressed. Nepal hopes that, sooner or later, India will listen to Nepal's concerns. The agreement on the Pancheswar project under the Mahakali treaty will be crucial, to both India and Nepal. This will be a unique project on the only large border river between the two countries.

The time has come for Nepal to approach India more meaningfully to work on a mutually beneficial agreement on multipurpose projects located wholly in Nepalese territory, wherein India buys not only power, but also cooperates in the integrated development of other benefits that come along. If these issues are ignored they will prove to be a big setback to the development process. In Nepal, the general opinion is that "no deal" is better than a bad deal. In the case of augmentation of low flow in the Ganges at the Farraka barrage (the lowest recorded flow of the Ganges at Farraka is 1,100 m^3/sec), the Kosi high dam would be an appropriate scheme because of its proximity to Farraka, and Nepal should seek access to the sea by developing a navigational channel from Nepalese territory. The Kosi high dam could be a multi-country venture in sharing the benefits it accrues in terms of power, irrigation, flood control, low-flow augmentation, and navigation.

Initiation of partnership-building efforts

After visualizing the stalemate in the government sector in harnessing the water resources of the South Asia region, an effort (supported by the Ford Foundation in 1990) was made by scholars of the region to forge partnerships between/among countries of the GBM basins. The Ford Foundation offered to support such an effort built around a core group of scholars from India, Bangladesh, and Nepal. The core group consisted of the CPR (Centre for Policy Research) in Delhi, the BUP (Bangladesh Unnayan Parishad) in Dhaka, and the IIDS (Institute of Integrated Development Studies) in Kathmandu.

The first phase of the partnership came to an end in 1994, culminating with the publication of three country reports and a consensus book, *Converting Water into Wealth* (Verghese *et al.* 1994). The book came up with an excellent resumé of possibilities for bilateral (and possibly trilateral) cooperation.

The second phase of the study started in February 1995 and is under way. It has so far conducted five dissemination/follow-up seminars, held in Dhaka in August 1995 and March/April 1997, in Delhi in October 1996 and November 1997, and in Kathmandu in January 1997. It planned to prepare a regional report in March 1998.

The objectives of the seminars were to increase the level of understanding among the countries of the region and facilitate the process of achieving regional cooperation in the development of water resources by exchanging ideas and information contained in national and regional reports. The effort has been successful in bringing together members of parliament, politicians, media people, and scholars of the co-basin countries in a common forum, and has been conducive to the movement towards regional cooperation in the development of the Himalayan water resources.

During the period of this cooperative study, the treaties between Nepal and India on the Mahakali and between India and Bangladesh on Ganges water sharing at Farraka, signed in the later half of 1996, have been landmark achievements and have encouraged the scholars to go ahead with the concept of regional cooperation.

Also, under the aegis of the United Nations Development Programme, a preparatory workshop for a project on enhancing regional cooperation in South Asia through cooperation on energy and environment was held in Singapore in December 1995. Policy-makers and senior specialists from India, Nepal, and Pakistan participated in the workshop, and formed working groups on natural gas, hydropower, energy efficiency and renewable energy, and hazardous waste management. These working groups identified a small number of topics on which specific proposals have been developed. The follow-up meeting of the group, the South Asia Forum, was held in Stockholm, Sweden, in July 1996.

Issues for detailed studies and research

There are many issues that need to be addressed further, not only for technical adequacy, but also to satisfy the public in the exercise of confidence-building. Some major issues of a technical nature are highlighted below, and are meant to be indicative rather than exhaustive.

River basins

Studies and deliberations held among scholars so far have clearly shown that unless scientific, technical, environmental, and socio-economic studies of the river basin as a whole are conducted, optimal utilization and sustainable exploitation of the water resources of the basin will not be achieved.

It is thus imperative that more rigorous basin, sub-basin, and micro-basin studies be carried out by each country, and the findings should be freely disseminated among national river basin institutions (to be established, if not in existence). A regional institution (such as the Ganges River Basin Authority) should be the custodian to collect, collate, package, retrieve, and disseminate the findings. This arrangement, however, should not be an impediment to the implementation of projects like Pancheswar and Karnali which are in advanced stages of development.

Hydrology and sedimentation

It has been surmised that the available hydrological data could be a thorny question when agreeing to certain technical parameters for a particular project. Fundamental questions like how much water is available, and when and where it is available, have still not been accurately answered. The temporal and spatial nature of monsoonal rainfall that prevails in the region has made it more difficult. The information available on hydrology is limited, with the quality of data varying. There is thus a need to analyse the data carefully and bring them up to a certain confidence level so as to be usable for project design.

The coverage of hydro-meteorological stations is inadequate, and to predict the trends, time-series historical data are needed; these are currently lacking. The available data are assessed to have less than 15 per cent reliability. An elaborate hydro-meteorological network and better management of the hydrological data should be given higher priority and monitored continuously on a regular basis through the process of ongoing research and analysis.

In storage projects, the major concern is the inflow of sediment mass and depletion of storage volume. The Ganges in general, and the Himalayan rivers like the Karnali, Gandaki, and Kosi in particular, are known to carry very high sediment loads. Continuous records of sediment loads of Nepalese rivers do not exist, but the total sediment load of Nepalese rivers from available data has been estimated to be 726 million imperial tons.

Any miscalculation in the estimation may lead to the premature sedimentation of reservoirs created by storage projects. The Kulekhani reservoir in Nepal is a classical example: in 1993 a cloudburst alone sedimented about 5 per cent (4.28 million m^3) of the reservoir capacity out of 83 million m^3 (designed value in 1982).

An elaborate network of stations on rivers of Nepal to collect sediment data should be established and research continued. At the same time, watershed studies and proper management should be given due importance in view of the river sedimentation.

Irrigation water management

It has been said many times and in many places that we need to practise economy in water use, to avoid wastage and get the most from a given amount of water. The irrigation sector is the largest consumer of water in South Asian society, and it is known that the irrigation efficiency at present in this region is less than 50 per cent. If the overall efficiency of water use can be increased, the issue of augmentation of low flow of the Ganges at Farakka may not arise at all. Hence, this aspect of water management needs more serious research, and best management practices should be looked for and propagated in the region. Continuous research and implementation efforts in this field are needed.

Environmental aspects

Storage projects generally have severe adverse environmental impacts in physical, biological, and socio-economic terms. The most significant impact is of course social, as large amounts of land are inundated and people have to be compulsorily resettled. This is one area which has gained considerable importance in recent times. In Nepal environmental impact assessment studies are mandatory in all hydro projects, and mitigation measures must be addressed as well. However, further research in this area is needed, and a more meaningful focus must be attached to it.

Dam safety and seismicity

The Himalayan belt encompassing Nepal is very active seismo-tectonically, and one of the most important design inputs is the design value of the seismic load in terms of the seismic coefficient. As dam safety is of utmost concern, rigorous studies and research should be conducted; more importantly, the analysis and results need to be transparent to the public at large as well (Sharma 1997).

There are many other issues which are of utmost importance, such as participation (of local communities in particular), rehabilitation of displaced persons, capacity-building of local and regional institutions, and inland water transportation and access to the sea for Nepal. These need further detailed studies/research and consideration at all levels of society. If inland water transportation is achieved, it will bring a sea-change to how Nepal does business with India and the rest of the world.

Need for a supranational institution

Now that several partnership-building efforts have been continuing for some time and knowledge-based information has been generated (much more work is needed, however), the time has come to agree on a broad regional water agreement among the Ganges river basin countries to continue, strengthen, and support such activities for the betterment of the people of this river basin. International agreements and conventions on major rivers of the world do exist as examples which can be studied and pondered upon. However, it is evident that a high-level (preferably ministerial level) apex body has to be constituted under such an agreement. The high-level apex body should have broad objectives, such as "to promote and coordinate effective planning and management for the equitable, efficient, and sustainable use of land, water, and environmental resources of the Ganges basin". There should be a mechanism whereby community participation is ensured, such as constituting a community advisory committee representing regional and special-interest groups, to give independent advice to the apex body on the views of the basin's communities.

The apex body could be named the Ganges River Basin Authority/Council. Its specific goals could be to:

- ensure sustainable development and utilization of water resources;
- maintain and improve, where possible, water quantity and quality for all beneficial uses;
- conserve the natural environment of the basin.

This high-level apex body should be supported by well-designed regional and national organizations. The apex body must be supported by an autonomous executive arm, called perhaps the Ganges River Basin Commission, with equal representation of the participating governments. The commission should have a secretariat of its own, consisting of technical and support staff in the areas of river management, natural resources, finance and administration, and communication. This commission should be able to interface with the existing regional organization, such as the SAARC.

The commission should have a knowledge-based approach. Hence, at the outset, its main function should be to collate, collect, package, retrieve, and disseminate the available data and knowledge as well as to generate more knowledge in all aspects of comprehensive river basin planning. These data and knowledge should be available and adopted to formulate strategies and policies. Each participating country should have a similar organization with similar objectives to interface and interact with the regional body. This is a crucial requirement and must be agreed to. This is one of the possible scenarios of organization which could be deliberated upon by member countries.

Good governance

One of the hurdles identified that has impeded the harnessing of the immense water resources of the region is good governance and the lack of political will. And, when it comes to implementing policies, programmes, or projects, it often depends upon the overall capacity of public administration. One cannot negate the existence of government and its accompanying bureaucracy. And it is the bottom line, period.

Governments should be forced to make good policies. However, getting the policies right may not, by itself, be sufficient for successful development. Good governance is required to ensure that governments actually deliver promises to their citizens. The governments of this region have failed miserably on this score, so far.

Well-designed programmes are important, but so is the ability to implement them. Implementation capability reflects three fundamental elements in the strategic interactions between citizens and government officials. These fundamental elements – accountability, transparency, and predictability – determine the performance of policy.

The 1990s showed some progress towards regional policies and co-operation in the region. The liberal policy towards its neighbours adopted by India (the so-called Guzral doctrine) and the democratization of polity of Nepal and Bangladesh auger well in this direction.

But good governance does not come by wishing for or enunciating policies alone. It has to be strived for and needs to be designed properly and honestly, and adhered to (legislation and rules should be enacted and applied in spirit as well as in letter). The culture of good governance has to be imbibed, and its fundamental elements have to be propagated.

Corrupt practices are rampant in the region, and must be shunned by the governments as well as society at large.

The people and governments of the region must realize that unless good governance is pursued and practised, the cherished goal of harnessing the water resources of the region could well be a day-dream.

Conclusions

The region consisting of the Ganges river basin is inhabited by some of the poorest people of the world, living amidst plenty of natural water resources which have not been harnessed to optimal and sustainable use. One of the main reasons is the lack of political will.

Isolated efforts in each of the countries are continuing, while the holistic and integrated river basin approach has eluded those striving for the optimal and sustained development of the water resources.

Past efforts have not been conducive to the balanced development of the resources, and have been the source of antagonism between the riparian countries. However, the efforts have resulted in some developments on some rivers after long lapses of time (more than 30 years in the case of Nepal and India).

Cooperation and partnership have been fostered on some fronts between/among nations to harness the water resources to uplift the unsatisfactory living conditions of the people of this region. The political atmosphere, at present, seems conducive to pursue such an effort.

Possible schemes to exploit the resources have been identified. Regulatory provisions seem to be evolving, partnership efforts among NGOs and at the academic level have been initiated, and consensus-building activities are continuing.

The time has come to establish a permanent supranational institutions(s) to promote and coordinate effective planning and management of water resources in the Ganges basin. A good governance system in each of the participating basin countries should be instigated. Partnership must be forged. The development of water resources of the region is inevitable and will remain a long-drawn-out affair; it might take as long as a millennium to achieve the desirable level of development, but we must start now.

REFERENCES

Bhadra, B. 1997. "Water Resources Development and Management Institutional and Legal Issues in a Regional Context", proceedings of International Conference on Large-Scale Water Resources Development in Developing Countries, Kathmandu.

Biswas, Asit K. and T. Hashimoto, eds. 1996. *Asian International Water, From Ganges-Brahmaputra to Mekong*, Water Resources Management Series 4. Sponsored by IWRA, with the support of UNEP and the UNU. New Delhi: Oxford University Press.

Char, N. V. V. 1997. "Integrated Water Resources Development of Ganga, Bramhaputra and Meghna River Systems – International Dimensions", paper presented to International Conference on Harnessing the Eastern Himalayan Rivers and Regional Cooperation, Delhi.

Hillel, Daniel. 1997. "Nepal Water Resources Strategy Formulation, Phase I Consolidated Draft Report", in consultation with B. K. Pradhan, team leader, HMG/Nepal/World Bank (J6F) Project.

Institute for Integrated Development Studies. 1995. *Water Resources Development – Nepalese Perspective*, Kathmandu: Institute for Integrated Development Studies.

Institute for Integrated Development Studies. 1997. "Regional Cooperation in Harnessing the Eastern Himalayan Rivers", briefs on topics, unpublished. Kathmandu.

Master Plan for Irrigation Development in Nepal. 1990. Cycle 2, CIWEC and EastConsult, Department of Irrigation.

Nepal Electricity Authority. 1996. "Medium Hydropower Studies Project."

Onta, P. R., N. K. Agrawal, and I. R. Onta. 1995. "National Irrigation Master Plan in Nepal", paper presented in the International Workshop on Inter Comparison of National Water Master Plans; sponsored by IWRA, UNEP, and Government of Mexico, March, Mexico City, Mexico.

Prasad, T. 1997. "Comprehensive Water Resources Development in the Indo-Nepal Region – Perspective and Prospects", proceedings of International Conference on Large-Scale Water Resources Development in Developing Countries, Kathmandu.

Sharma, C. K. 1997. "Limiting Factors for Dam Designing in Nepal", proceedings of International Conference on Large-Scale Water Resources Development in Developing Countries, Kathmandu.

Shrestha, Hari M. and Lekh M. Singh. 1996. "The Ganges-Brahmaputra System: A Nepalese Perspective in the Context of Regional Cooperation", in *Asian International Waters, From Ganges-Brahmaputra to Mekong*, eds Asit K. Biswas and T. Hashimoto. New Delhi: Oxford University Press, pp. 81–94.

Verghese, B. G. *et al.*, eds. 1994. *Converting Water into Wealth – Regional Cooperation in Harnessing the Eastern Himalayan Rivers.* Centre for Policy Research.

7

Water resources development of Nepal: A regional perspective

Ratneshwar Lal Kayastha

Introduction

Nepal is rich in water, and so is the Ganges basin. The world's average water availability is 269,000 m^3/km^2, whereas water availability in the Ganges basin is 334,000 m^3/km^2 and that of Nepal is 1,182,000 m^3/km^2 (Table 7.1). But despite having four times more water availability than the world's average, Nepal has been able to irrigate only 50 per cent of its irrigable land, and only 60 per cent of the population has easy access to drinking water.

More diversions to provide year-round irrigation water will necessitate tremendous investment. Small and medium-size rivers have already been diverted for irrigation. Major rivers such as the Kosi, the Gandak, and the Mahakali have been diverted in the past, and agreements between India and Nepal to facilitate this were also concluded, but those diversions are mainly irrigating Indian land. Similarly, the Karnali has also been diverted by India constructing a barrage in Indian territory. To meet the year-round irrigation demand in Nepalese territory, new diversion schemes have to be implemented on all big rivers; on the one hand, these will require tremendous amounts of financial resources which at present are beyond Nepal's internal capacity, and on the other hand, for those structures being built on transboundary rivers, external investors will seek transboundary understandings before implementation. Calculation shows that Nepal requires only one-third of its total water and two-thirds

Table 7.1 Water availability in the Ganges basin

	Land surface (km^2)	Runoff (km^3/annum)	Water availability (m^3/km^2/annum)
Earth	148,900,000	40,000	269,000
Ganges	1,087,300	363 (at Hardinge Bridge)	334,000
Left tributaries of the Ganges			
Nepal and Tibet	180,700	208	1,107,000
Nepal	147,181	174	1,182,000

will flow downstream even after fulfilling the country's whole requirement, but the basic fact is that Nepal is not able to irrigate more than half of its irrigable land by diverting the abundant available surface flows. This syndrome means Nepal will have to develop its groundwater to irrigate the remaining irrigable terai land over the next 15 years on a conjunctive use pattern. It may be a similar situation in India and Bangladesh when year-round irrigation requirements have to be addressed in the Ganges basin area. But if a certain flow discharge is required for maintaining draft in the river for navigational purposes and also for controlling salinity in the coastal area, the situation becomes worse as regards available water for irrigation and other consumptive uses. So not only Nepal but all the partners of the Ganges basin are rich in water resources and poor in its utilization. To overcome this requires enormous financial resources, transparency, understanding, and combined willingness; only then will the Ganges basin, which is rich in water resources, become rich through its utilization.

Water availability

Water availability in Nepal has its own patterns, which have implications for the Ganges basin. The patterns are mainly due to its geographical location, high Himalayan mountain ranges, and seasonal variations. An abundance of water at one time becomes a scarcity at another time. To regulate these variations is a Herculean task. Total annual runoff from all the rivers flowing in and through Nepal is 208 bcm. The Ganges basin generates 502 bcm. The average annual runoff of the Ganges basin is 8,727 m^3/ha (Verghese and Iyer 1994) of cultivable land and of Nepal is 77,700 m^3/ha. Monsoon flow (June to October) accounts for 85 per cent of the total runoff. Reservoir sites identified in Nepal account for 77 bcm of live storage capacity which will generate an additional flow of 4,950 cubic metres per second in the dry season (December to May).

Hydrology

Nepal's hydrology can be analysed in terms of river basin flows and groundwater reserves. Precipitation occurs in the form of rainfall and snowfall. Snowfall trapped in the high Himalayas augments the dry-season flows by melting, hence its contribution is accounted for in the Himalayan river flows. Permanent snow lies above 4,500 metres and accounts for 10 per cent of the total precipitation of the country. There is a wide variation of precipitation from east to west. The average annual rainfall is about 1,500 mm in a good monsoon regime. Depending upon the variations of precipitation around the country, flood peaks vary over time.

River basins

The quantum of water available in Nepal has to be analysed on the river basin approach, taking into account river flow characteristics. Rivers are grouped into three categories depending upon their origin and seasonal discharge variations: Himalayan rivers; Mahabharat rivers; and Churia rivers.

Himalayan rivers

Himalayan rivers are snowfed rivers and their dry-season flows are augmented by snow melts. The snow-capped Himalayas work as a reservoir and release their water in the dry season. High dry-season flows are needed for irrigation and other consumptive purposes, but only 15 per cent of the total amount of water is available during the dry season in comparison to 85 per cent of water available in the monsoon season. But the cost of diversion structures on these rivers is tremendous and hence it has not been possible to utilize the dry-weather flows to their maximum in Nepalese territory. The rivers that fall into this group are the Kosi, Gandaki, Karnali, and Mahakali. The ratios of maximum monthly flows to minimum monthly flows in the Kosi, Gandaki, Karnali, and Mahakali rivers are 15:5, 19:0, 14:3, and 11:2 respectively. If instantaneous flows are taken into account the ratios will be much more. In the case of the Karnali river it is about 130:1. Hence, the construction cost of the diversion structure will be many times more than the cost of diversion structures which can be constructed on regulated rivers. Construction of reservoirs in upstream reaches will drastically decrease the maximum design discharge. The total annual reserving capacities of the Mahakali, Karnali,

Gandaki, and Kosi river basins are estimated to be 7 bcm, 34 bcm, 18 bcm, and 14 bcm respectively. The mean monthly flows of Himalayan rivers are shown in Table 7.2.

The discharge in the monsoon season from June to October accounts for 82 per cent of the whole quantum of flow. Dry-weather flow, although appreciable for Nepalese consumptive requirements, is difficult to utilize. These rivers contribute significantly to the dry-weather flow of the Ganges river. High dams constructed on these rivers will augment the Ganges water by 170 per cent in the dry season, and lower riparian countries would benefit tremendously. The annual water volume flowing from these rivers is 176 bcm, which is 85 per cent of the total water available from Nepalese territory.

Mahabharat rivers

The Mahabharat rivers originating from the Mahabharat hills (mid hills) are not snowfed, hence dry-season flow is basically due to groundwater recession. The main rivers in this category are the Kankai, Kamala, Bagmati, West Rapti, and Babai. The total stream flow is 18 bcm, of which 87 per cent flows in the monsoon season (June–October). The percentage of dry-weather flow is less compared to Himalayan rivers. The basin areas of these rivers, especially the river valleys, are thickly populated and hence there are no feasible storage sites except on the Kankai. The water from these rivers contribute 9 per cent of the annual Nepalese flow. Table 7.3 shows the average monthly water availability from Mahabharat rivers.

Churia rivers

There are numerous seasonal rivers originating from the Churia hills which flow in Nepal or through to India. Due to the highly permeable character of the catchment area, dry-season flow becomes non-existent in many of these rivers, and only a few have a small amount of water flowing in the dry season. The total volume of water available in these rivers is estimated to be about 14 bcm, out of which 91 per cent flows in the monsoon season. Flood discharges are flashy in nature, and carry heavy silt during the monsoon period. There are no reservoir sites on these rivers, so monsoon flows in excess of water utilized for irrigation purposes cannot be stored for future utilization. Table 7.4 shows the volume of water available from the Churia rivers, which contribute only 6 per cent to the annual Nepalese flow. The dry-season flows where available and part of the monsoon flows have already been utilized traditionally by the farmers. The main fertile terai land lies in the catchment areas of these rivers.

The available monthly flow from all these rivers combined is shown in Table 7.5. The flow available during the monsoon season from June

Table 7.2 Mean monthly flows of Himalayan rivers (million cubic metres)

River	January	February	March	April	May	June	July	August	September	October	November	December	Annual
Mahakali	514	433	458	540	817	1,665	4,852	4,091	4,426	1,773	676	607	20,852
Karnali	1,015	832	954	1,177	1,928	4,046	9,106	11,884	7,960	3,576	1,672	1,224	45,374
Narayani (Gandaki)	1,063	784	801	1,016	1,679	4,720	12,883	14,903	10,052	4,786	2,317	1,450	56,454
Kosi	1,074	837	921	1,185	2,041	4,710	11,999	12,937	9,272	4,382	2,250	1,497	53,105
Total monthly flow	3,666	2,886	3,134	3,918	6,465	15,141	38,840	43,815	31,710	14,517	6,915	4,778	175,785
Monsoon flow from June to October									144,023	(82% of annual discharge)			(85%)

Table 7.3 Mean monthly flows of Mahabharat rivers (million cubic metres)

January	February	March	April	May	June	July	August	September	October	November	December	Annual
250	204	186	176	294	1,604	4,520	4,696	3,619	1,505	564	352	17,970
Monsoon flow from June to October									15,944	(87% of annual discharge)		

Table 7.4 Mean monthly flow of Churia rivers (million cubic metres)

January	February	March	April	May	June	July	August	September	October	November	December	Annual
228	143	128	110	123	765	3,058	5,654	2,369	1,297	386	259	14,520
Monsoon flow from June to October									13,143	(91% of annual discharge)		

Table 7.5 Available monthly flow from all Nepalese rivers (million cubic metres)

January	February	March	April	May	June	July	August	September	October	November	December	Annual
4,144	3,233	3,448	4,204	6,882	17,510	46,418	54,165	37,698	17,319	7,865	5,389	208,275
Monsoon flow from June to October									173,110	(84% of annual discharge)		

to October accounts for 84 per cent of the total annual available discharge.

Groundwater

Groundwater is mainly available in the south of Nepal (the terai area), which is part of the Ganges alluvial plain. The extraction of groundwater in mountainous regions is not technically and economically feasible. The annual groundwater availability is 11,598 million cubic metres, as shown in Table 7.6; 76 per cent of its volume can be extracted for consumptive purposes. This volume will meet 14 per cent of the irrigation requirements of the terai area. Since groundwater contributes to the dry-season river discharge, extraction of groundwater means diminishing the dry-season flow in the river. These environmental considerations have an impact on the level of groundwater extraction. To keep the groundwater table within the limits, so that no waterlogging is created in the irrigable area and also a minimum river discharge is maintained for ecological purposes, proper limitations have to be imposed. Similarly, groundwater extractions have to be limited to allow salinity control in the coastal area. As groundwater contributes to the river flow, in terms of total water availability it cannot be counted separately.

Water requirements

Water requirements have to be analysed for both the present pattern and future projections. The present methodologies adopted for conveyance systems and usage patterns may not be applicable in the future, and present uses cannot be directly related to the future patterns. For example, currently open unlined channels may be lined in the future, or distribution systems at field levels may be through buried pipes. Similarly, domestic as well as industrial water requirements may change depending upon future consumption patterns. The total projected annual water requirement for future use is 64 bcm.

Irrigation

Irrigation requirements constitute the largest percentage of the whole consumptive requirement. As the irrigation water requirement depends upon cropping patterns, presently cultivation of paddy is consuming the greatest share. The cultivation of paddy, wheat, maize, oilseeds, vegetables, sugarcane, and jute, which are dominant crops at present, will also continue in future. The future requirement for year-round irrigation

Table 7.6 Recharge of shallow and deep aquifers in terai areas of Nepal

Zone	Gross area (km^2)	Estimated area of *bhabar* (km^2)	Surface area (km^2)	Arable land for irrigation (km^2)	Recharge to *bhabar* (million m^3)	Recharge shallow aquifer (million m^3)	Total recharge (million m^3)
Mahakali	1,273	125	1,148	1,121	80	421	501
Seti	1,876	244	1,632	1,599	157	646	803
Bheri	2,528	369	2,159	2,109	165	492	657
Rapti	2,995	437	2,558	783	144	842	986
Lumbini	4,841	625	4,216	3,162	681	1,994	2,675
Narayani	5,032	743	4,289	3,693	433	1,222	1,655
Janakpur	2,870	394	2,476	2,213	241	532	773
Sagarmatha	2,687	475	2,212	1,712	304	721	1,025
Koshi	2,648	344	2,304	1,889	296	1,028	1,324
Mechi	1,716	258	1,458	1,219	260	939	1,199
Total				19,500	2,761	8,837	11,598

Table 7.7 Irrigation water requirements – Nepal

Terai				
	East (area 981,851 ha)		West (area 761,513 ha)	
Months	Diversion requirement per 1,000 ha (million m^3)	Volume required (million m^3)	Diversion requirement per 1,000 ha (million m^3)	Volume required (million m^3)
January	1.477	1,450	1.612	1,228
February	1.775	1,743	1.868	1,422
March	3.061	3,005	2.934	2,234
April	4.406	4,326	2.443	1,860
May	3.277	3,218	1.076	819
June	1.524	1,496	1.600	1,218
July	1.819	1,786	1.295	986
August	2.932	2,879	1.468	1,118
September	1.529	1,501	1.363	1,038
October	2.589	2,542	2.405	1,831
November	2.219	2,179	2.403	1,830
December	1.254	1,231	1.394	1,062
Total		27,356		16,646
Annual requirement				44,002

Hill				
	East (area 187,547 ha)		West (area 186,201 ha)	
Months	Diversion requirement per 1,000 ha (million m^3)	Volume required (million m^3)	Diversion requirement per 1,000 ha (million m^3)	Volume required (million m^3)
January	1.294	243	1.410	263
February	1.598	300	1.518	283
March	3.219	604	2.625	489
April	3.080	578	2.115	394
May	1.783	334	0.860	160
June	4.496	843	3.610	672
July	5.969	1,119	5.668	1,055
August	5.278	990	6.070	1,130
September	5.923	1,111	5.019	935
October	7.611	1,427	6.212	1,157
November	3.694	693	3.985	742
December	0.577	108	0.557	104
Total		8,350		7,384
Annual requirement				15,734

Table 7.7 (cont.)

	Mountain			
	East (area 32,905 ha)		West (area 28,133 ha)	
Months	Diversion requirement per 1,000 ha (million m^3)	Volume required (million m^3)	Diversion requirement per 1,000 ha (million m^3)	Volume required (million m^3)
January	1.189	39	0.979	28
February	1.908	63	1.380	39
March	2.294	75	1.618	46
April	0.911	30	0.737	21
May	2.342	77	1.740	49
June	6.118	201	4.989	140
July	5.569	183	5.244	148
August	4.693	154	5.498	155
September	5.857	193	4.939	139
October	5.588	184	4.465	126
November	2.673	88	2.713	76
December	1.225	40	0.918	26
Total		1,327		993
Annual requirement				2,320

Source: Canadian International Water and Energy Consultants/East Consult 1989

will be 62 bcm annually, which represents 98 per cent of the total water requirement. Hence, if savings are to be made, cropping patterns and existing methodologies for water conveyance and field application have to be addressed. This will certainly require more investment. To make it economically viable the price of produce also has to be investigated.

The total irrigation water is not lost: 50 per cent of water flows into the groundwater reserves as a return flow and augments the river flows, but this occurs mainly on the downstream side. Volumes can only be evaluated by a total basin water-balance study, which is of great importance when assessing the total consumptive demand of the basin. The future water requirements for irrigation, based on ecological regions and total water requirement, are given in Tables 7.7 and 7.8. The total irrigable land available by region and by ecological belt is shown in Tables 7.9 and 7.10. Comparison reveals that numerous water sources are available in the mountains lying in the catchment of Himalayan rivers, whereas most of the irrigable land lies in the catchment of the Churia rivers where there is a scarcity of water.

Table 7.8 Total monthly irrigation water requirement (million m^3)

Ecological belt	January	February	March	April	May	June	July	August	September	October	November	December	Total
Terai	2,678	3,165	5,239	6,186	4,037	2,714	2,772	3,997	2,539	4,373	4,009	2,293	44,002
Hill	506	583	1,093	972	494	1,515	2,174	2,120	2,046	2,584	1,435	212	15,734
Mountain	67	102	121	51	126	341	331	309	332	310	164	66	2,320
Total	3,251	3,850	6,453	7,209	4,657	4,570	5,277	6,426	4,917	7,267	5,608	2,571	62,056

Source: Canadian International Water and Energy Consultants/East Consult 1989

Table 7.9 Irrigable area by region – Nepal

Region	Terai (ha)	Hill (ha)	Mountain (ha)
Eastern	442,262	77,533	14,760
Central	539,589	110,014	18,145
Western	282,712	99,084	280
Mid-western	245,679	50,962	12,480
Far western	233,122	36,155	15,373
Total	1,743,364	373,748	61,038
Grand total			2,178,150

Source: Canadian International Water and Energy Consultants/East Consult 1989

Table 7.10 Irrigable area by ecological belt – Nepal

	Eastern (ha)	Western (ha)	Total (ha)
Terai	981,851	761,513	1,743,364
Hill	187,547	186,201	373,748
Mountain	32,905	28,133	61,038

Source: Canadian International Water and Energy Consultants/East Consult 1989

Domestic, commercial, and industrial

The future water requirement for drinking purposes and other commercial and industrial uses, according to recent reports, is 1.4 bcm annually, which is nearly 2.3 per cent of future irrigation requirements (Tables 7.11 and 7.12). Though domestic water requirements seem to be small in comparison to future water availability, the hard fact is that there is an acute scarcity of water in hill and mountain areas, and also in Kathmandu valley, mainly because the population concentration is away from the available water sources. It will require great effort and investment to provide easy access to potable drinking water to the future population.

Navigation

Navigation is quite important for Nepal's economy as transportation costs are drastically reduced if goods are transported via water routes. Being a landlocked country, having navigational routes would be a lifeline for Nepal. India has already declared that the Ganges will be a navigable river from the sea up to Allahabad. Nepal can be connected to the Ganges along the Kosi, Gandak, and Karnali rivers for navigational pur-

Table 7.11 Projection of rural daily domestic water requirement for 2026

Region	Population (000)	Domestic (m^3)	Livestock (m^3)	Students (m^3)	Health centre (m^3)	Healthpost (m^3)	Total (m^3)
Eastern	5,915	354,900	70,980	11,830	112	3,914	441,736
Central	9,253	555,180	111,036	16,822	133	6,547	689,718
Western	6,485	389,100	77,820	16,216	112	5,028	488,276
Mid-western	4,019	241,140	48,228	6,927	105	3,059	299,459
Far western	2,009	120,540	24,108	2,513	63	1,534	148,758
Total							2,067,947
Monthly future water demand		62.9 million m^3					
Annual future water demand		755.0 million m^3					

Source: Sharma 1996

poses. Though river routes, being natural routes, incur less cost for land acquisition and lower construction costs, river cross-sections which have stabilized to accommodate high flood discharge generate less draft during dry-weather flows and bigger ships cannot move. Hence, to maintain greater depth in the river a higher flow is needed. One way to minimize river flow and maintain draft is by providing locks as and where appropriate. A second alternative can be a controlled separate navigational channel, as mentioned in earlier literature. A non-consumptive water-use alternative should be discussed rather than a flowing water alternative required for maintaining the river draft. More joint studies are needed to reach an agreeable solution.

Salinity control

Salinity can be viewed as two separate problems, one due to use of fertilizers and pesticides in irrigated areas, and the other arising in coastal areas. In Nepal, high monsoon flows and comparatively steeper ground slopes accelerate the leaching effect of the salt and hence excess salinity is not a noticeable problem. In the Ganges basin, salinity has to be addressed in the coastal areas. The main countermeasure is to maintain a particular groundwater level and slope. This necessitates a controlled extraction of groundwater during dry weather, and recharging groundwater by diverting water for irrigation uses and also by maintaining a certain water level in the rivers near coastal areas. Maintaining water depth by water flow during the dry season demands high volumes of water, hence serious research work must be done to arrive at an economical and amicable solution.

Water requirement – a regional perspective

Spatial and temporal variation

From the graph shown in Figure 7.1 it is evident that the available water concentration is in monsoon season. Diverting water which is available in Himalayan rivers involves high construction costs, hence rivers have not been diverted for the maximum benefit of the irrigable land in Nepal. Diversions have been built in the Kosi as part of the Sunsari-Morang irrigation project and the Kosi pump canal scheme (irrigated area 71,000 ha). From the Gandak river 40,000 ha of land is being irrigated; similarly, the Mahakali river irrigates 11,000 ha of land. The terai area, which has 1.7 million ha of irrigable land (80 per cent of total land), can get water for irrigation only from the Mahabharat and Churia rivers, which cannot

Table 7.12 Projection of urban and industrial daily water requirement

Town	1996 pop.	Demand (million litres/day)	2001 pop.	Demand (million litres/day)	2011 pop.	Demand (million litres/day)	2026 pop.	Demand (million litres/day)
Kathmandu	562,981	93.2	752,383	127.1	1,343,785	237.2	3,207,495	599.3
Biratnagar	151,838	25.1	178,184	30.1	245,382	43.3	396,555	74.1
Lalitpur	139,411	23.1	167,742	28.3	242,846	42.9	423,023	79
Pokhara	135,895	22.5	193,811	32.7	394,210	69.6	1,143,539	213.7
Birganj	86,850	14.4	109,309	18.5	173,154	30.6	345,219	64.5
Dharan	83,643	13.8	105,273	17.8	166,760	29.4	332,472	62.1
Bharatpur	76,808	12.7	107,912	18.2	213,005	37.6	590,704	110.4
Mahendranagar	73,917	12.2	88,053	14.9	124,953	22.1	211,229	39.5
Bhaktapur	69,234	11.5	78,061	13.2	99,235	17.5	142,237	26.6
Janakpur	68,515	11.3	85,802	14.5	134,565	23.8	264,290	49.4
Hetauda	67,084	11.1	83,592	14.1	129,793	22.9	251,123	46.9
Dhangadi	57,464	9.5	73,785	12.5	121,651	21.5	257,536	48.1
Nepalganj	56,680	9.4	67,183	11.3	94,389	16.7	157,185	29.4
Damak	53,430	8.8	69,087	11.7	115,512	20.4	249,728	46.7
Butwal	52,739	8.7	62,825	10.6	89,153	15.7	150,709	28.2
Bhairawa	52,228	8.6	69,104	11.7	120,979	21.4	280,232	52.4
Tribhuvanagar	34,433	5.7	40,814	7.7	57,341	10.1	95,490	17.8
Birendranagar	29,646	4.9	38,257	6.5	63,709	11.2	136,909	25.6
Rajbiraj	29,443	4.9	35,783	6	52,851	9.3	94,866	17.7
Tulsipur	26,986	4.5	32,148	5.4	45,620	8.1	77,118	14.4
Inarwa	23,180	3.8	28,971	4.9	45,255	8	88,350	16.5
Gaur	22,799	3.8	25,437	4.3	31,665	5.6	43,978	8.2
Lahan	22,430	3.7	26,453	4.5	36,796	6.5	60,364	11.3
Byas	22,041	3.6	24,141	4.1	28,960	5.1	38,050	7.1
Kalaiya	20,525	3.4	22,775	3.8	28,041	5	38,308	7.2
Taulihawa	20,442	3.4	24,400	4.1	34,765	6.1	59,122	11
Bidur	19,692	3.3	20,743	3.5	23,016	4.1	26,902	5

Jaleshwor	19,274	3.2	20,537	3.5	23,319	4.1	28,212	5.3
Dhankuta	18,963	3.1	21,063	3.6	25,984	4.6	35,605	6.7
Bhadrapur	18,953	3.1	23,617	4	36,670	6.5	70,948	13.3
Malangwa	16,162	2.7	18,470	3.1	24,123	4.3	36,005	6.7
Ilam	15,333	2.5	17,814	3	24,047	4.2	37,712	7
Dipayal	14,090	2.3	16,062	2.7	20,873	3.7	30,922	5.8
Tansen	13,874	2.3	14,154	2.4	14,732	2.6	15,643	2.9
Banepa	13,731	2.3	15,040	2.5	18,042	3.2	23,705	4.4
Dhulikhel	11,630	1.9	13,785	2.3	19,368	3.4	32,253	6
Total	2,202,344	364.3	2,772,570	469.1	4,464,549	788.3	9,473,738	1,770.2
Monthly future water demand			53.84 million m^3					
Annual future water demand			646 million m^3					

Source: Bhattarai 1996

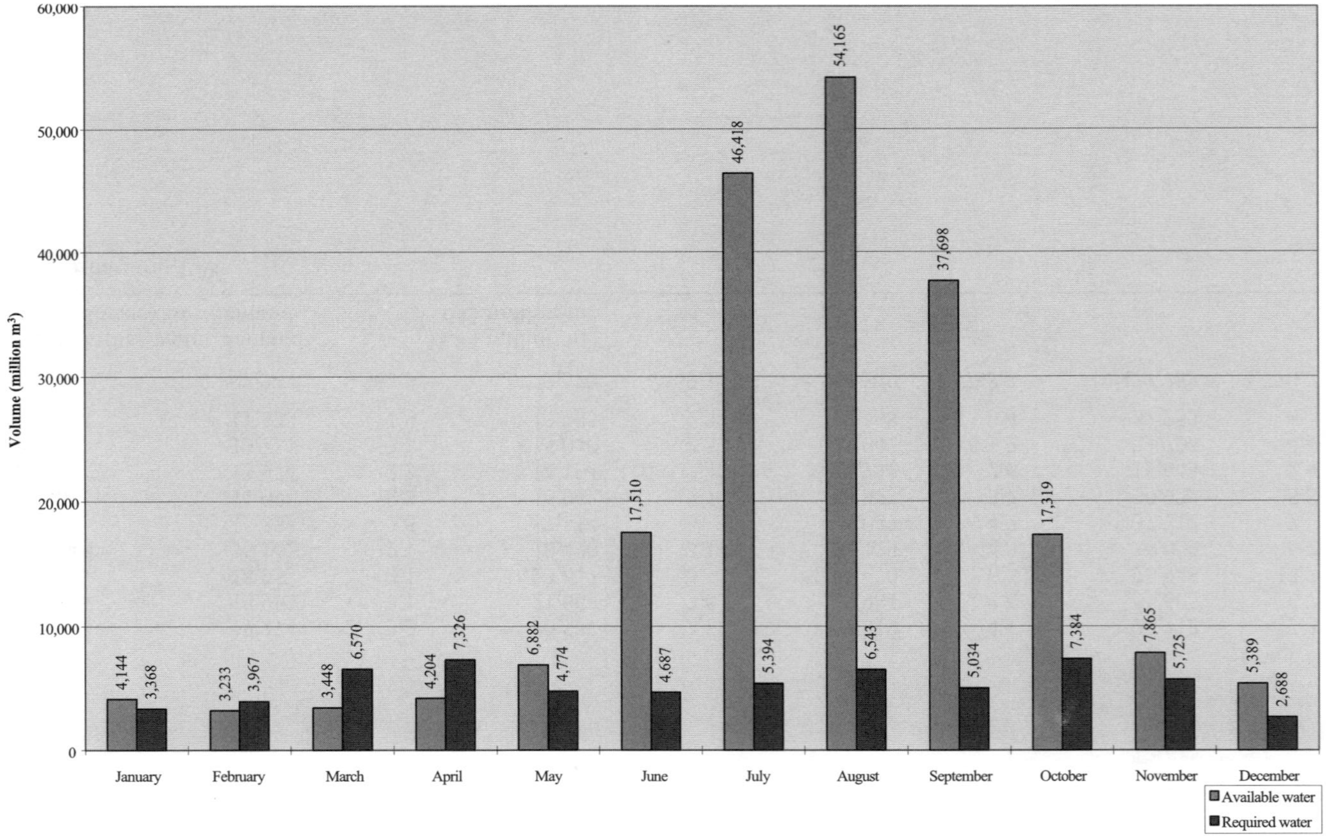

Figure 7.1 Monthly water requirement versus availability

supply year-round irrigation. Only about 50 per cent of presently irrigable land receives year-round irrigation. Almost all the dry-season water of the Mahabharat and Churia rivers is utilized in Nepalese territory, hence water downstream by surface flow becomes scarce. India has invested considerably in irrigation projects, and diversion structures have been constructed on many small rivers where there is a scarcity of water in the dry season. It would have been wise to have a joint dialogue before investing in these projects. If one major project with storage facilities had materialized, this would have been far more beneficial than having many small projects without water. Even the huge investment in the Kosi and Gandak projects, if analysed economically, will raise many questions. Regulation of flow is a must for optimization of net benefits.

Like seasonal variations of water flow, spatial variation of availability is imposing serious problems. In the Kathmandu valley round-the-clock water supply has not been possible and at present only 71 per cent of inhabitants get a water supply. In the hills many people have to walk for hours to fetch water for their daily requirements. Hill valleys are situated at such an elevation that gravitational water diversion imposes serious technical and economical problems. Simpler water diversion schemes, whether in the hills or the terai, have already been developed by the farmers themselves. Only diversion schemes which are technically more difficult and require more investment remain to be implemented. In the hills diversion of water for irrigation purposes has an adverse effect on slope stability. The abundance of water is thus not translated into an abundance of available water. Spatial and temporal variations in the availability of required water impose great constraints on overall development activities, especially that of agriculture, which is the major source of GDP.

Here it is appropriate to mention that when studying future demand for consumptive water for Nepal, it is easy to see that the dry-weather flows of Himalayan rivers are almost totally required for utilization within Nepal itself. The cropping pattern envisaged in the master plan for irrigation development in Nepal (Canadian International Water and Energy Consultants/East Consult 1989) shows a scarcity of water during February, March, and April of almost 40 per cent (see Figure 7.1), thus even Nepal's future requirement cannot be fulfilled by the available water in those months. This will certainly necessitate storage projects upstream.

Value-added augmentation

Any augmentation of the natural available water flow will require substantial investment. Investment in the present context of the world economy will require maximization of net benefits. For maximization of net benefits, basin development on a regional basis becomes inevitable. Country boundaries vanish and the river basin becomes a single unit. Lessons can be learned from the past. Major projects have been constructed where benefits were limited within a single country boundary; had those been developed on a maximization-of-benefits principle, maximum return would have been realized in terms of hydropower, irrigation, navigation, and flood control. But this cannot be achieved unless the countries involved have a will to work together and share all the achievable benefits accruing from a project. This will certainty involve sharing the cost in proportion to the benefits accruing, and every cumec of augmented flow will require the sharing of costs between the countries which utilize the water. But evaluation of benefits is very difficult. All the regional partners have to agree on the principle of evaluating the benefits, sharing the benefits, and sharing the cost.

Constraints

Augmentation of flow means creation of reservoirs in suitable locations. Nepal has very suitable reservoir sites which can store about 77 bcm of water and generate an average flow of 4,950 m^3/sec in the dry season from December to May. A tremendous amount of benefit will accrue from this. But at what cost? Cost does not only include investment in physical facilities cost, but also the hazards associated with the development of the project, such as submergence, population displacement, increase in seismicity, ecological changes, and other environmental impacts. All hazards associated with these projects are borne by the country where the development takes place, and a proper mechanism should be formulated to account for them. Not only that, but natural resources have to be valued according to the availability of alternatives in the market for those resources, so that a proper justification can be given for valuing the benefits arising out of the resource. When we talk about augmentation of flow, catchment improvement schemes also have to be considered. Presently, the water carries thousands of tonnes of sediment through the rivers into the sea, which has an adverse effect on the storage capacity of the reservoirs, channel capacity, river erosion, navigational characteristics of the rivers, and other factors. Catchment improvement thus has to be thought of as a part of any reservoir project.

Diversification of water utilization

Present water use, whether for irrigation, navigation, or coastal salinity control, has to be readdressed. For consumptive use, irrigation water is the main requirement. Cropping patterns can be readdressed depending upon the future food grain requirement and the possibility of export and import of food grain to and from other countries. The efficiency of irrigation can be improved by providing lined channels and buried-pipe distribution systems, sprinkler irrigation, and other methods of providing irrigation water.

For navigation use, a network of navigational waterways, whether through rivers or channels, has to be considered. Instead of maintaining draft by flowing water, navigational locks are an alternative. This will certainly require huge investment, but the possibility cannot be ignored. Detailed joint study with an exchange of data will be needed.

For salinity control in the coastal areas, the fresh water front has to be maintained at certain level depending upon the sea water levels. River locks to maintain fresh water fronts can be a solution. A study has to be conducted. Similarly, for controlling the advancement of the saline front, groundwater extraction has to be controlled. Upstream groundwater extraction may not have any impact on the saline fresh water front, but will certainly decrease the dry-weather river flow.

Waste water can be treated and reutilized, which will be better for environmental purposes. Reutilization can be considered for irrigation purposes.

Power development

The hydropower potential of Nepal has been estimated to be about 83,000 MW, out of which 42,000 MW has been identified as economically feasible. Hydropower development is associated with the development of multipurpose projects which also generate irrigation, flood control, and navigation benefits. The flow augmentation from multipurpose projects will meet the dry-season demand of Nepal and will also augment the flow of the Ganges. The electricity demand of Nepal by 2025 is projected to be nearly 1,700 MW, so there will be a vast surplus which can be made available in the region to multiply the growth rate in several sectors, including industry, agriculture, and tourism.

Hydropower should be visualized as an eco-friendly energy source which also provides peaking energy. It is a replenishable source of energy which, if utilized, will keep in reserve the finite energy sources such as coal and gas.

In the regional concept, power generation and high-voltage transmission lines can be planned, so that eco-friendly hydropower can be transmitted in the region to stimulate economic growth as a whole. Sitting on this vast potential source of energy without utilizing it will hinder future growth. This region, one of the poorest in the world, will remain so in future unless the countries involved work together.

Multipurpose hydropower projects need huge investment which is difficult to generate from within the region (Nepal, India, Bangladesh, and Bhutan), so private developers have to be called for. Private developers are mainly interested in selling electricity and seeking returns on energy sales. Other benefits such as irrigation, flood control, and navigation are not their immediate concern. So to achieve these benefits all the riparian countries have to agree, and a benefits-sharing principle should be worked out accordingly. Unless this is done, electricity will not be available at a cheaper rate and the full potential benefits cannot be realized.

Multiple trade-offs

Benefits from multipurpose water resources development projects should not only be shared on their own virtues, but can also be traded for other benefits needed by each riparian country. For example, Nepal needs navigational routes to sea and simpler trade and transit facilities. Indian needs more hydropower and water for its consumptive use, mostly for irrigation purposes. The vast fertile lands of the Ganges plain require huge amounts of water for year-round irrigation and inland navigation. Similarly, Bangladesh requires water for irrigation, navigation, and salinity control of coastal areas. So multimodal benefits should be shared according to individual countries' needs. This approach will help the countries concerned in their overall economic growth.

Regional understanding – a compulsion and not an option

Analysing the overall developments of water resources in the Ganges basin, there can be no option other than having a strong, effective, trustworthy regional understanding among the riparian countries. It calls for basin development and not development within individual countries. As has already been explained, there is an abundance of water from the total availability point of view, but at the same time there is temporal scarcity. In the past there has been acrimony over sharing water in the dry season when demand for water in the basin exceeded its availability. During the monsoon season when there is abundance, the heavy floods are detrimental to all parties. So there is a problem around the year, and in the present age of technological development regional understanding has become a compulsion and not an option. Future generations cannot be sustained without resolving these issues. Doubling of the population, water requirements, food grain requirements, and the deterioration of the environment in the coming decades are threatening us; we must act now or prepare for mass destruction. So regional cooperation and understanding is a must, and we cannot say "no". Such cooperation can be achieved at two levels: political and bureaucratic.

Politicians or people's representatives are the policy-makers. They are aware of the people's hardship and of national priorities. They should be made aware of the realities and policy frameworks should be agreed upon by high-level politicians. This policy guidance should be specific up to workable limits. But the desires of high-level politicians do not necessarily translate into actions. Policy matters unresolved at bureaucratic levels should be taken care of at political levels. Politicians of every side should have the logic behind the arguments of other sides explained to them, so that they talk openly.

Bureaucrats know the details of the issues more clearly and in more depth than politicians. In the past, negotiations have been held mostly at bureaucratic level, and have concentrated on theoretical discussions on minor issues without a practical outlook. Had this not been the case, we would not have been talking about these vast potential water resources for over four decades while there continued to be acute scarcity of water and energy. Whatever water diversion treaties were agreed upon were not accepted by people at large as balanced treaties. But now the realization has come. Treaties on the Mahakali river and Farakka barrage have shown positive signs that all the partners feel comfortable. But the bureaucratic hangover is still there, and few things are yet clarified. But people need the water resources potential to be exploited, as without this living conditions will deteriorate day by day and stagnation of economic growth may take place due to the increasing population. So an institution has to be created where overall Ganges basin development matters can be discussed, data exchanged, research and studies conducted, and further policies enumerated. At present discussions on the hydrological data, water availability, and effects of dry-weather flows are all developed by single sides and have not generated full trust. Even the hydrological data which are available are not very dependable and require further refinement. To develop trust among the parties, exchange of data, confirmation of data, and analysis of data are very important, otherwise doubt will continue. So let us create a common forum, discuss common goals, and achieve these jointly.

REFERENCES

Bhattarai, Kiran K. 1996. *Water Supply and Sanitation (Urban)*. Water Resources Strategy Formulation Programme (Phase I). Kathmandu: Ministry of Water Resources.

Canadian International Water and Energy Consultants/East Consult. 1989. *Master Plan for Irrigation Development in Nepal*. Planning and Design Strengthening Project. Kathmandu: Department of Irrigation.

Sharma, S. N. 1996. *Water Supply and Sanitation (Rural)*. Water Resources Strategy Formulation Programme (Phase I). Kathmandu: Ministry of Water Resources.

Thapa, Bhekh B. and Bharat B. Pradhan. 1995. *Water Resources Development, Nepalese Perspective*. Delhi: Konarak Publishers.

Verghese, B. G. and R. R. Iyer. 1994. *Converting Water into Wealth – Regional Cooperation in Harnessing the Eastern Himalayan Rivers*. Delhi: Konarak Publishers.

8

Water resource development and the environment in Bhutan

Somnath Mukherjee

Introduction

Small is beautiful, and Bhutan demonstrates this. Half of the size of Arunachal Pradesh, Bhutan exudes charm. Its magnetic mountains, virgin forests, delightful people, and crystalline air invoke surrealism. Stretching from lush humid forests in the south to dry alpine steppe in the north, Bhutan is home to some of the best remaining representatives of Himalayan wildlife and habitat, and is considered one of the global "hot spots".

Buddhism has imbued a strong sense of respect for all forms of life, and thus conservation is not a new concept to Bhutan. But modern development is, at least in the way most people conceive of it. Surrounded by developing countries with desires for material luxuries, Bhutan cannot keep itself aloof from the process of development. But development needs money, and the small country of Bhutan, with limited resources, has little to sell except timber, hydropower, and tourism to raise the major part of the finance required for development. Bhutan, however, will have to tread cautiously in the new century to balance the development of environmental resources and the development process itself to make it sustainable.

This chapter makes an attempt to assess the potential of water resource development in Bhutan, and its environmental implications in the perception of the author.

Table 8.1 Catchment and subcatchments in Bhutan

Catchment	Subcatchment	Area (km^2)
Torsa		3,486.40
	Jaldhaka	1,021.35
	Torsa	2,465.05
Raidak		4,954.03
	Ha	976.69
	Paro	1,040.82
	Wongchu	1,365.80
	Raidak	1,242.50
Sankosh		10,434.16
Mao		1,856.28
Mangde Chu		75,522.63
	Mangdechu	4,444.20
	Bumthang	3,079.43
Manas		9,562.08
	Kuru	4,052.86
	Sheri	378.20
	Khulanchu	1,806.38
	Gawri	1,162.80
	Manas	1,771.94
	Diuri	390.50
Nyera		1,568.86
Dhansiri		865.56

Source: Negi 1983

The water resource scenario

The whole of Bhutan forms a part of the Brahmaputra basin. The main rivers forming the watershed catchment of Bhutan are the Torsa, Raidak, Sankosh, Mao, Manas, Nyera, and Dhansiri. The largest catchment area is that of the Sankosh. Table 8.1 shows the catchment and subcatchment areas of rivers, and Figure 8.1 shows the catchments on a map.

Except the Mao, Nyera, and Dhansiri, all the rivers representing the major catchments of Bhutan receive water from the high Himalayan glaciers. In recent times the phenomenon of global warming has been particularly evident in Bhutan and has caused enormous damage to the environment, life, and property. Due to global warming, the glacial snouts recede, leaving large quantities of morainic material and water. The water often gets bunded by a morainic "dam". As the quantity of water increases due to further melting of glaciers, the morainic dams burst, resulting in the sudden release of large quantities of water into the nearby tributaries. The water flows down with great ferocity, causing havoc downstream. The Sankosh in particular is susceptible to such glacial bursts. During a glacial burst in 1994, hundreds of thousands of trees got

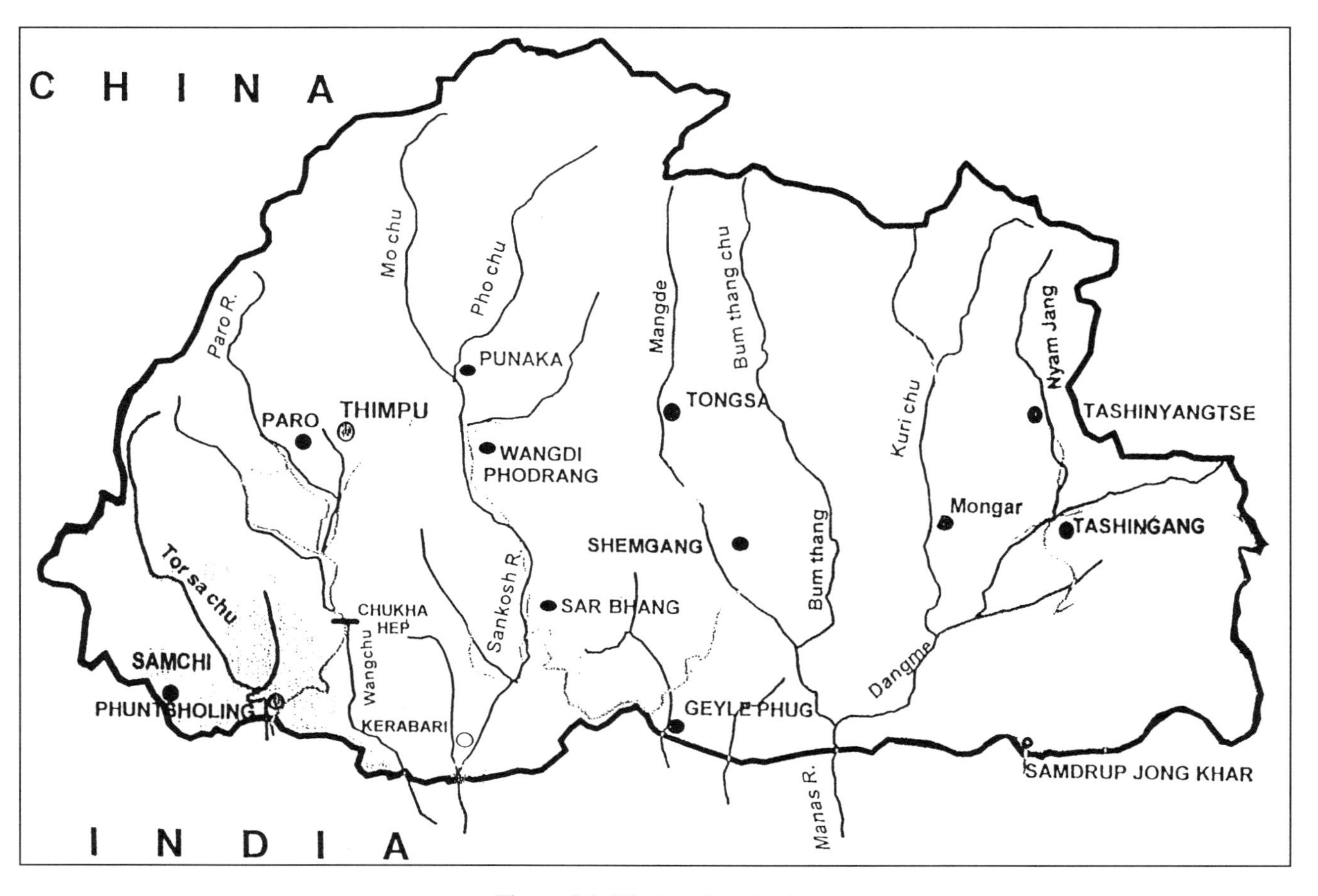

Figure 8.1 Bhutan river basins

washed away, innumerable cattle were killed, human life was lost, large tracts of agricultural fields along the valleys were engulfed, and part of Punakha *dzong* (district), probably the most sacred *dzong* in Bhutan, was destroyed.

The groundwater potential of Bhutan has not been investigated. There was some effort with Japanese assistance to estimate the groundwater potential in southern Bhutan at the beginning of the 1990s. A few test wells were planned, but had to be abandoned due to socio-political problems. However, the rural areas rely heavily on springs for domestic and irrigation use.

The quality of surface water is generally excellent. Due to higher gradients and cascades the DO (dissolved oxygen) levels are extremely high. As the population is scanty and scattered, the point source of untreated domestic effluent is negligible and therefore the BOD (biochemical oxygen demand) is minimal.

Bhutan does not have many large industries, and therefore the problem of water pollution due to effluents does not arise. Farmers use little or, in most cases, no fertilizer or pesticide, and water is free from pollutants that are commonly encountered in agricultural areas.

The majority of the rivers which originate in Bhutan flow through India and transverse through Bangladesh. The path Bhutan adopts for water resource development thus has implications in India and Bangladesh with respect to quantity and quality of available water.

Water-use potential

Availability of water is not a constraint in Bhutan – it is a country of springs, rapids, and rivers. But the basic problem relates to harnessing the water for various uses. The population density of Bhutan is extremely low and scattered. A small and scattered population means that the per capita cost of delivery of the resource to consumers is extremely high. In spite of this, Bhutan has done reasonably well in the water resource sector.

Most impressive is probably the drinking-water supply: the majority of the rural population have a piped water supply. Spring water is stored at a higher elevation in a cement tank, and water is supplied through gravity to the consumers by pipes. Since the source is at a higher elevation and mostly in remote locations, contamination of water is rare; only a few incidents of faecal contamination of the piped water supply have been reported. The most notable and recent water contamination happened in Chukha in 1994, when faecal contamination at the source resulted in an outbreak of cholera in the area. But with few exceptions, the quality and quality of water supply is reasonably good and the way the water supply system is being managed needs to be appreciated.

Table 8.2 Major land-use categories in Bhutan

Land-use category	Area (km²)	%
Perpetual snow	4,006.30	10.0
Barren exposed rocky areas and scrubland	3,762.64	8.7
Water spreads	522.92	1.3
Agricultural land, including settlements and orchards	3,558.74	8.8
Forests, including alpine pastures, and temperate and tropical forests	24,290.37	67.0
Degraded forest	1,415.90	3.0
Plantations	27.38	0.1

Source: Negi 1983

Major land-use categories in Bhutan are shown in Table 8.2. About 8.8 per cent of the land in Bhutan is under agricultural use (including orchards and settlements). The data (Dorji 1992) show that about 25 per cent of the cultivable land is irrigated and devoted mainly to paddy cultivation. Another 14 per cent of the cultivable land in under orchards, and about 1 per cent constitutes kitchen gardens. Shifting cultivation constitutes a prominent proportion and accounts for about 30 per cent of the agricultural land, occuring mainly in the rainfed areas. Per capita land and forest distribution in rural areas are shown in Table 8.3.

All over Bhutan, agricultural lands are scattered through small settlements. Terraced land mostly produces paddy. Much permanently used rainfed land is often not terraced, and therefore the use of water is not optimal. However, lack of water, as such, is not a constraint. What is a constraint is water management in terms of irrigation development and associated development of land by terracing.

Because of the geomorphologic characteristics, Bhutan has a good hydropower potential. The first hydro-electric power plant of 360 kW was commissioned in 1967 at Samtelingchu in Thimpu valley. So far only about 1.7 per cent of the total estimated theoretical potential of 20,000 MW has been harnessed (Tamang 1993). Bhutan at present has 11 hydro-electric stations with an installed capacity of 341.75 MW, out of which Chukha station alone generates 336 MW. The existing stations are shown in Figure 8.2. Apart from these stations Bhutan also has 11 mini-hydro stations which generate about 0.4 MW. The power master plan preliminary study (1991) listed 89 sites from a desktop study of topographical maps, with a potential of approximately 12,000 MW. Further, the power system master plan (PSMP) report of 1992 listed 24 sites which are economically feasible and have a cumulative identified potential of 10,988

Table 8.3 Per capita land and forest distribution in rural areas

Dzongkhad	Per capita (ha)	
	Land	Forest
Chirang	0.90	0.63
Dagana	3.62	3.10
Gasa	32.78	12.17
Gaylegphu	1.88	1.58
HA	11.43	2.67
Jakar	13.89	7.30
Luntshi	8.08	5.28
Mongar	2.89	2.50
Paro	5.05	4.07
Pensagatshel	1.47	1.00
Samchi	1.25	0.89
Samdrupjongkha	3.25	2.50
Shemgong	6.00	5.39
Tashigang	2.53	1.91
Thimpu	4.04	3.39
Tongsa	7.43	6.24
Wangdisphordang	9.90	6.73
Average	3.97	2.80

Source: Condensed from Negi 1983

MW. Most of the schemes are run-of-the-river type, with the exception of a few reservoir schemes, mainly along the southern border. The proposed sites in the PSMP report are shown in Figure 8.3. However, the PSMP figure needs revision, as the envisaged capacity of quite a few identified projects has been increased. For example, the Tala project is being constructed with an ultimate capacity of 1,025 MW instead of 265 MW as shown in Figure 8.2. Similarly, the Sankosh project has also undergone an upward revision to 4,000 MW instead of 376 MW. It is thus clear that the PSMP study needs to be updated and the economically feasible power projects can probably generate more than 15,372 MW.

Whatever the revised PSMP figure might be, it is evident that Bhutan has a hydropower potential of much more than 250 kW per square kilometre. Tamang (1993) calculated that the total household requirement of electricity for the whole of Bhutan is about 161 MW, considering a family size of six and dwelling space of 50 square metres per family. This means that if Bhutan decides not to go for dramatic industrial development, it has an power export potential of more than 15,200 MW. On the other hand, Bhutan's immediate neighbour, India, is a power-starved country. Though at present the eastern grid of India does not have much power demand, the gap between generation and demand is likely to

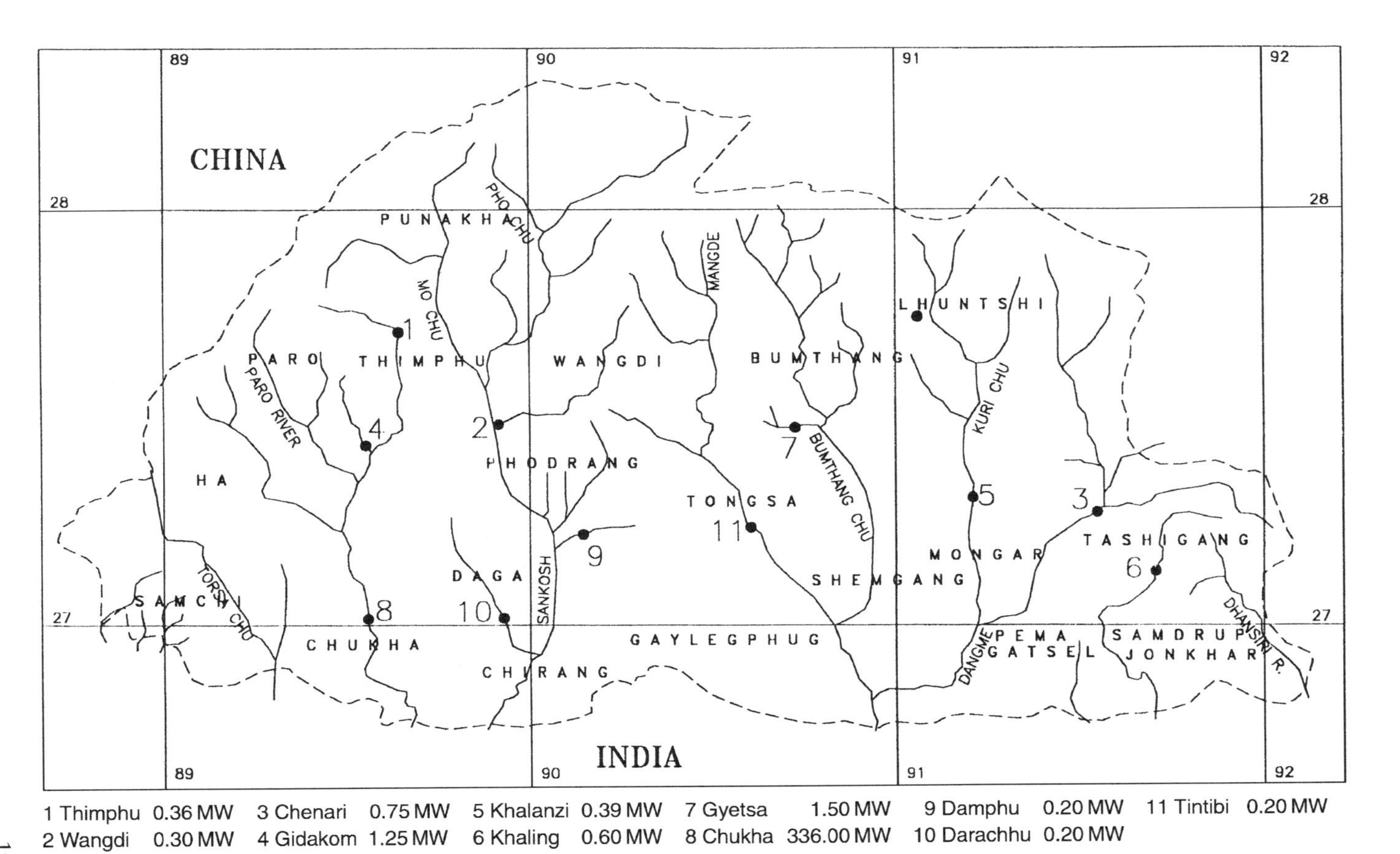

Figure 8.2 Bhutan – existing hydro-electric power plants

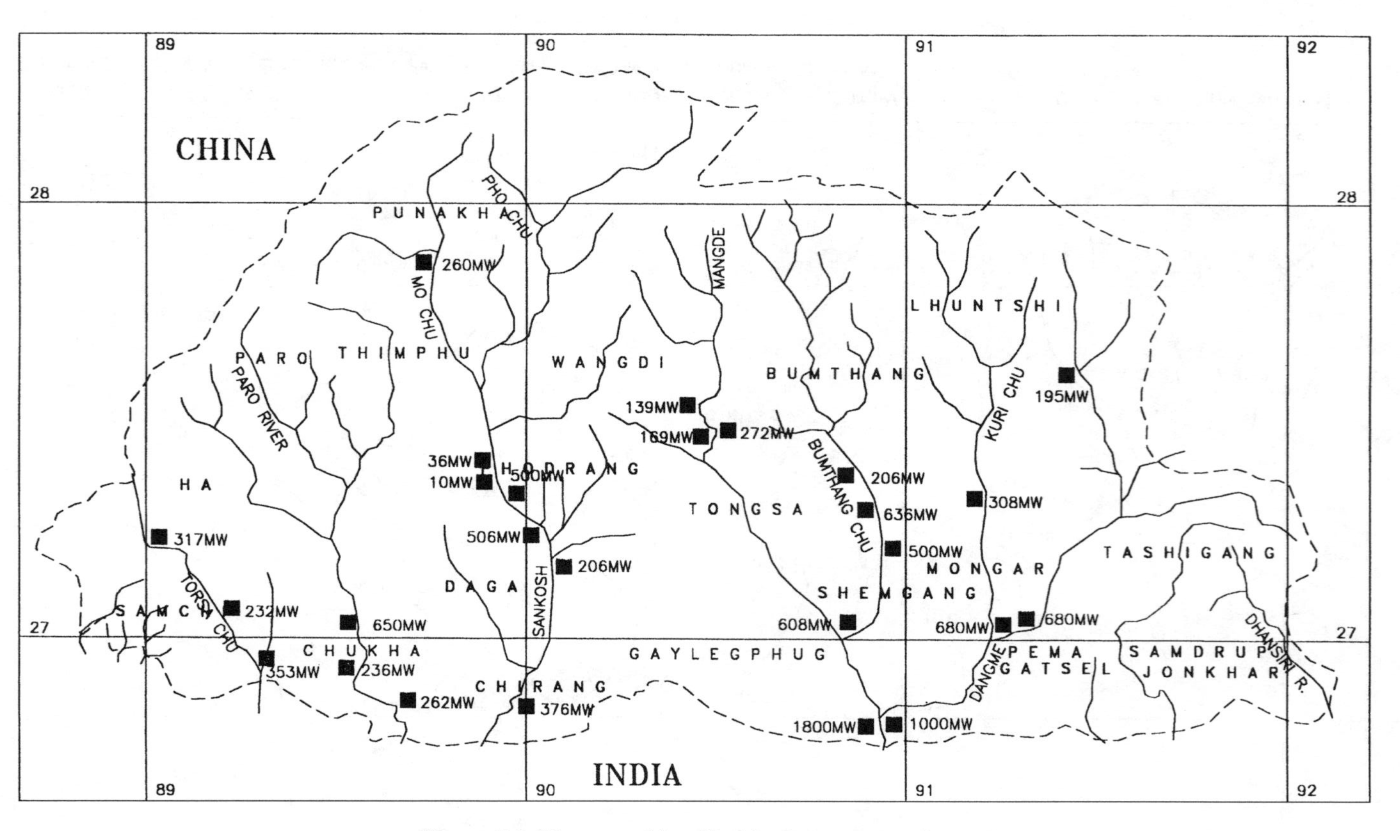

Figure 8.3 Bhutan – identified hydropower projects

widen as further industrial development occurs. Moreover, even if the demand in the eastern grid does not increase, the demand in other grids is predicted to increase drastically. Since the grids are connected, power from Bhutan in the eastern grid can flow to various parts of India.

The environmental scenario

Bhutan is one of the 10 global "hot spots" identified by the United Nations. Geographically, Bhutan combines two bio-geographic zones, namely the Palearctic and the Indo-Malayan, which account for the wide range of flora and fauna in the kingdom. With an altitude of 150 metres to more than 7,000 metres above sea level, the climate varies widely from warm and humid subtropical to permafrost types.

The eco-floristic zones of Bhutan are shown in Figure 8.4. The border between India and Bhutan contains probably one of the world's best examples of moist tropical forests. However, migration of population from India, Nepal, and Bangladesh threatens a large tract of prolific forests, especially in the Samchi and Chirag areas. Two of the bio-reserve areas of Bhutan are partly located along the lush, moist tropical zones. The moist subtropical ecofloristic zones are immediately adjacent to the band of moist tropical regions to the north. The subtropical forests exist mainly along the Sankosh, Tongsa, Kurichu, and Dangme valleys. Montane forests can be seen in the high hills of southern and central Bhutan, whereas alpine pastures are found in the northern high hills.

From Table 8.2, it is apparent that forests cover 67 per cent of the land in Bhutan. Table 8.4 gives the percentage of forest and non-forest land along the watershed catchments of Bhutan. Amongst the major watershed catchments, Manas has the largest tract of forest land (ignoring Raidak and Mao, which are small), whereas Sankosh has the maximum percentage of non-forest land. In the Punaka area (Sankosh valley), large-scale deforestation has taken place to make way for human habitation.

The faunal diversity of Bhutan is also substantial. The high hills are the home of takin, bluesheep, wolf, and clouded leopard. The southern portion supports tropical species like Asiatic elephants, one-horned rhino, tiger, etc. Golden langoor, one of the rare primates, are found in Bhutan. Though poaching is against the traditional ethics of Bhutanese people, of late it is common along the Indian border. Problems of militancy have increased the poaching in this area, as the market for wild animals harvests easy money without much risk of penal actions.

Bhutan has an impressive 23.2 per cent of its area declared as protected. These protected areas cover varied eco-floristic zones, and their management is being improved. However, lack of funds and experience

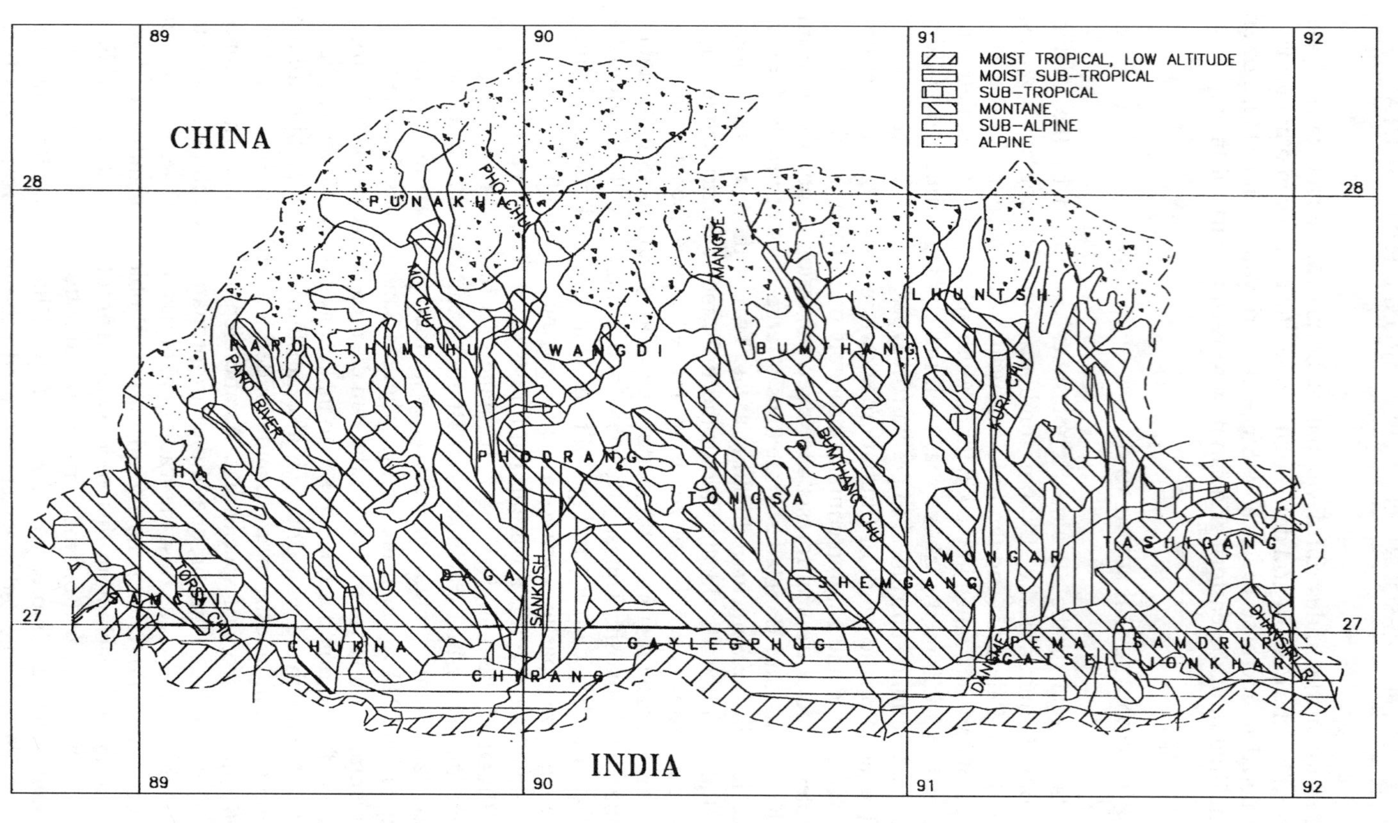

Figure 8.4 Bhutan – eco-floristic zones

Table 8.4 Forest and non-forest land by watershed catchment

Catchment	Non-forest (%)	Forested (%)	Total (%)
Torsa	24.8	75.2	100
Raidak	20.4	79.6	100
Sankosh	39.0	61.0	100
Mao	13.3	86.7	100
Mangdechu	33.9	66.1	100
Manas	27.1	72.9	100
Nyera	21.9	78.1	100
Dhansiri	20.4	79.6	100

Source: Negi 1983

of modern management systems leaves a large gap between the desired and the achieved goals.

Schiotz (1993) believes that Bhutan has taken a rather hesitant stand in the field of surveying its biodiversity. This is probably due to the feeling that there are higher priorities for the government and that such surveys can wait until Bhutan has its own trained ecologists. But it needs to be appreciated that time is probably running out for Bhutan, and a long-term programme could be developed with some appropriate and knowledgeable expatriate assistance.

Cost of development

Need for development

For centuries Bhutan has isolated itself from the rest of the world. The reason for this (geographical inaccessibility, self-imposed isolation, or political necessity) is not a point of argument. When the curtain lifts, we find a unique country with a rich natural heritage, traditional agricultural practices which allow nutrient preservation, and a people with a profound respect for nature. Bhutan, as many believe, could possibly provide a model for the world in nature conservation (Schiotz 1993).

The basic question here is, with such an ideal status, why should Bhutan develop? The country has sustained itself and its people for ages; why cannot Bhutan sustain itself for a few more centuries in the same way it has done in past? But high rates of population growth and political compulsions mean that Bhutan will probably have to adopt a development path which is different to the country's traditional practices.

Reliable data are not available on the precise annual population growth rate in Bhutan. Figures available vary from 2 per cent to more than 3 per cent per year. The land-use planning project (LUPP), which is responsible for generation of data and land-use planning (under the Ministry of Agriculture), considers 3.1 per cent as a reasonable annual population growth rate and uses the same for planning purposes. A population growth of 3.1 per cent per year means that the population will double in 23 years. If Bhutan wants to maintain its status quo on food production and continue traditional agricultural practices, the agricultural land must be doubled in the next 23 years. This is an impossible task given the morphological characteristics of the Himalayas and Bhutan's present economy. At a population growth rate of 3.1 per cent, there will be an additional 98,900 persons in the next five years. Bhutan would need additional revenue to provide education, jobs, shelter, etc. to the next generation of Bhutanese. No country with a population growth rate of 3 per cent per annum has so far been able to maintain a rising standard of living for its people and have surplus to a planned use of its renewable resources. The only option Bhutan has is to generate additional financial resources through exports. The harnessed power from water resources can generate the additional revenue required. In fact, Bhutan is doing this at present: a major portion of Bhutan's revenue comes from export of power from Chukha.

The other reason Bhutan has to develop relates to the philosophy of its neighbour. India is opening its economy to the world. It is debatable if indigenous technology or knowledge improves because of such globalization; but it is certain that consumption patterns will change and consumerism will become evident. With the attractions of consumerism so close to hand, Bhutan may find it difficult to resist the temptation. According to Bhutanese tradition, the person who consumes the least is respected the most. Such a concept is bound to change in future. Bhutan cannot isolate itself as an island of ascetics. It will be a political compulsion to develop to meet the aspirations of its people.

Environmental cost and benefit of water resource development

The argument has been put forward above that it is necessary for Bhutan to generate additional revenue to keep the country going, and water resources, in the form of hydropower, are probably the only exportable commodity which can maintain a high GNP to offset the strains of population growth.

The technology in hydropower generation has improved over the years, and efficient turbines are available which can generate electricity with less damage to the ecology. Overlaying the eco-floristic zone map with the proposed hydropower station map would suggest that the proposed projects are spread over large and ecologically diverse areas.

Some of the feasible projects are located in highly ecologically sensitive areas.

Bhutan at present has forest cover of 70 per cent. Very few countries of the world can boast of such high forest cover, despite the fact that a sizeable part of northern Bhutan is under permanent snow cover. The rural Bhutanese have taken the environment for granted, and believe that the vast stretches of forest land will be always available for exploitation. This may be true at present in view of the low population density, but in the experience of the author most Bhutanese are not aware that with a higher population growth rate, the forest resources would be under pressure and may not be so easily available for exploitation. The élite clan is well aware of the problem, but often not aware of the solutions. A brighter aspect is that the environmental awareness of an average administrator or engineer in Bhutan is greater than in other developing countries; but they need more training to arrive at the best possible solution to the problems.

Due to the development of water resource projects, Bhutan is bound to lose some of its highly biodiverse areas, especially in the south where some large projects have been envisaged. The loss of forest due to a hydropower project is not restricted to the submerged area only – a significant amount of forest gets affected by factors like construction activity, accessibility of otherwise remote areas, concentration of population around the project site, etc. To understand the exact amount of loss and the benefits of a project is a difficult task. The forests in Bhutan have not been evaluated with respect to species diversity. The unspoiled ecosystem constitutes a reservoir of genetic material of unknown value, but we know that it cannot be replaced once it has been destroyed.

Apart from the terrestrial flora and fauna, the other victim of hydropower development would be the fish. Most of the Bhutan rivers contain a variety of fish that play an important role in providing adequate nutrition to the rural population. Some of these fish are migratory in nature, and will be affected due to the construction of dams across the river: they would not be able to migrate for spawning, especially where the dam heights are significant and fish ladders cannot be provided. Masheer (*tor tor*) is an example. However, some of the important migratory fish, like trout, could be protected with significant efforts as they can survive in colder waters and can migrate to upper segments of the river.

The rehabilitation and resettlement problems are much less in Bhutan compared to other countries in this region which are densely populated. Due to the small and scattered population, the displacement of population caused by projects in Bhutan is low. But it must be understood that even the displacement of a relatively small population could be a problem for a small country like Bhutan where the options for rehabilitation and resettlement are extremely limited.

This probably gives a depressing scenario of water resource development, but the intention is not to underplay the benefits of development. Availability of power will not only generate financial resources but will also help in environmental conservation. The per capita fuelwood requirement of Bhutan is extremely high, and a good amount is used for heating in both rural and urban areas. Availability of electricity will substantially reduce the wood requirement for heating, thereby saving forests. Electricity will also help Bhutan to tackle the problem of achieving higher food production in future. The majority of Bhutan's rivers flow through gorges, and therefore the water cannot be used for irrigation. Availability of electricity will enable the country to pump water from the deep gorges to irrigate the fields. Groundwater can also be pumped and utilized for irrigation. The advantage of having electricity has thus never been questioned; what is important is how to trade off between having it and not having it.

Conflicts of conservation

Conservation in a developing country is often considered an élitist ideal with few beneficiaries. The fact is, in a poor country a vast population desperately depends on the exploitation of nature for fuel, food, and fodder. A poor population is concerned about today and has little inclination to think for tomorrow. Bhutan is no exception. Conservation imposes restrictions on forest-related activities which directly affect the economy of the rural poor. It is rather optimistic to think that poor people will cooperate with the government in forest conservation unless alternative means of income are provided to them to augment the loss. Emphasis on conservation is likely to bring friction between the government and the population who depend on forest resources for sustenance.

It is evident from past experience that the impact of degradation of environment in one part of the world is not restricted to the immediate vicinity but could affect faraway places. For example, Bhutan represents a sensitive ecosystem, and if large-scale deforestation takes place in the upper reaches of Bhutan, massive erosion may be a prominent phenomenon. A significant amount of silt from Bhutanese rivers will be deposited in India and Bangladesh, causing devastating floods. For this reason, many well-meaning environmentalists consider that a pristine environment is not only an asset to the country to which it belongs but is also an asset to the world. A sizeable amount of finance is available for environmental protection in the form of loans or donations from international agencies. But many view the lenders and donors with suspicion. Loans often come with a tag which is not acceptable to many, so accepting large loans or donations for environmental conservation could create internal political problems which Bhutan may not like to face at present.

In many developed countries environmental protection rights are vested with trusts or even individual persons. Theodore Panayotou at Harvard University proposed a system in Bhutan where the right to develop an area could be sold to and bought by individuals in the country or abroad through transferable development rights (TDR). TDR will find a market value depending not only on the market value of the potential development but also on the environmental value of the area. Developing countries often view such concepts with caution. In Bhutan, or indeed in the whole region, such concepts may not be acceptable to a large section of the population.

Options for Bhutan

To develop or not to develop could be a question. But there is little doubt that Bhutan will have to develop, and on the same lines, willingly or unwillingly, as all other developing countries develop. The question is, what are the options available to Bhutan for environmentally sustainable development?

There are three sectors of development through which Bhutan could generate financial resources: timber export; power export; and tourism.

Timber exports directly affect the ecosystem and are often not renewable. Moreover, for a profitable timber business there is always a tendency to plant exotic species which are fast-growing and have greater timber value. Such species are, however, extremely harmful to the ecology. The local faunal character often gets totally altered or even eliminated due to planting exotic species. Timber exports, however, cannot be totally discontinued under the present circumstancess, but should be kept at a low priority. Bhutan must undertake a review of the domestic use of timber. Most houses in rural areas use timber and mud as construction materials. The use of wood is not optimum, and a significant quantity is wasted. Logical design and better tools can prevent this wastage. In addition, the per capita use of fuelwood in Bhutan is too large. In areas producing *ara* (a local wine), consumption is as high as 6.7 kg per day per person. It was estimated by Raina (1992) that by the year 2000 the country's fuel consumption would increase up to 2,800 million kg. This means that about 400 ha of land must be denuded each year to meet the fuel requirement. With such a high rate of domestic consumption, the export of timber on a long-term basis cannot be sustainable.

Power export appears to be the most favourable option to generate financial resources for development. If the site and sizing of the project are judiciously selected, the ecological damage could be minimized. However, it must be realized that power projects have long gestation periods and are capital intensive. The projects will have to be planned well in advance to accrue benefits in time. The real demand for export and the capacity of the transmission system to evacuate power will have to be properly assessed before undertaking a project.

Tourism has become a major industry, especially for countries which are endowed with vast natural wealth. However, it can be culturally disruptive, especially where the tourists come from cultures which are very different from that of the country being toured. For that reason Bhutan has a most cautious tourist policy and has avoided the erosion of its cultural values (Schiotz 1993). Under the existing system the same policy is likely to continue, and therefore tourism has restricted potential.

From the above, it is concluded that power export has the maximum sustainable development potential. However, Bhutan has its own philosophy. The country may not want a quantum leap from the present state to a richness beyond imagination. Chukha has provided some comforts to the Bhutanese economy. Tala, a project about three times bigger than Chukha, is under construction and may provide surplus in the immediate future. Bhutan may like to wait and watch the impact of the surplus money before developing other projects. It may like to chew the amount it can digest rather than suffer from the obesity of electro-dollars. But certain measures could be adopted to ensure that the hydropower projects can operate without significant degradation to the environment.

The basic need for sustainable development is probably awareness: awareness of the fact that what is sustainable today may not be sustainable tomorrow. The whole environmental system must be appreciated in proper perspective. In this part of the world the developers and the environmentalists are always in confrontation. Appreciation of each other's point of view is totally lacking and the communication gap is often too wide to bridge. What both interest groups forget is that unless the development is sustainable, it is not development. Though the appreciation of environment in Bhutan is reasonable, much remains to be done through an appropriate curriculum at all levels.

The real genetic wealth of Bhutan is yet to be assessed scientifically. All we know is that it is enormous. A systematic study must be conducted to understand the genetic value of each area, so that the selection of project sites can be undertaken more logically. Based on such a study, it would be easier of designate areas which could be sacrificed for development purposes. Systematic data must be generated.

However, for a country like Bhutan, with widespread and biodiverse areas, whichever site is chosen for development some loss of genetic material will be inevitable. To restrict the loss of genes, the concept of a gene bank could be considered. Under expert guidance, the genetic material which needs protection could be selected and stored for future use. In several parts of the world, including a few in India, such gene banks are operative.

Bhutan is in need of trained manpower. Since there is no university in Bhutan, most students are trained abroad and import foreign concepts which in some cases may not be applicable in the Bhutanese context. It is necessary to have in-house trained personnel who can utilize traditional available knowledge combined with modern techniques.

In addition, the following measures also need due consideration:

- clear indications of environmental losses;
- rehabilitation of degraded areas;
- a proper and sound watershed management system;
- development of the necessary environmental legislation;
- respect for and maintainance of indigenous lifestyles.

Conclusion

Bhutan has a large population growth rate and, due to morphological characteristics, the land resources are limited. To feed, educate, and provide other infrastructural facilities for the increasing population, Bhutan needs additional finances. It is desirable that Bhutan generates its own resources to meet these financial demands rather than depending on international loans, donations, etc. Export of power to its neighbouring countries appears to be a feasible option. To make the export sustainable, Bhutan will have to care for its environment. Bhutan has been wisely maintaining its environment in a very sustainable manner in the past, and has profound traditional knowledge about the environment. However, in the changing scenario this traditional knowledge may need augmentation.

REFERENCES

Dorji, Pem. 1992. "The Agriculture Sector Policy: An Emphasis on Conservation and Sustainability," in *Bhutan and its Natural Resources*. Delhi: Vikas Publishing.

Negi, G. S. 1983. *Forestry Development in Bhutan: Report on Remote Sensing Land and Vegetation Mapping*. Project under the UNDP and FAO.

Panayotou, T. Undated. *Conservation of Bio-diversity and Economic Development*. Harvard Institute of International Development.

Raina, B. N. 1992. "Energy, Environment and Bhutan," in *Bhutan and its Natural Resources*. Delhi: Vikas Publishing.
Schiotz, A. 1993. *Nature Conservation in Bhutan. A Background Paper on Bhutan's National Environmental Strategy*. Bhutan: National Environmental Commission, Thimpu.
Tamang, B. 1993. *Hydropower Development in Bhutan: A Background Paper on Bhutan's National Environmental Strategy*. Bhutan: National Environmental Commission, Thimpu.

9

From dispute to dialogue to doing

B. G. Verghese

The Ganges-Brahmaputra-Meghna/Barak (GBM) system, child of the monsoon, cradled in the Himalayas, constitutes one of the world's greatest natural resources. This would be reason enough to harness its immense potential. But draining as it does a demographically exploding region, already home to over half a billion of the world's poorest people, makes this imperative. After decades of agonizing inter-regional disputation that stalled progress, a window of opportunity has fortunately opened at last.

These mighty rivers have long been used as waterways of commerce, and the Ganges has been diverted over the centuries to irrigate the rich alluvial Gangetic plain in India. The Brahmaputra and Barak/Meghna, however, remain largely untouched. Nepal and Bhutan, long secluded in the folds of the Himalayas, have more recently awakened to the need for developing their water resources, primarily to generate plentiful, cheap, clean energy. Nepal also has a need for irrigation and flood moderation. These requirements could be met by Himalayan storages, which would also mitigate flooding and augment lean-season flows in the Indo-Bangladesh plains below to stabilize and enlarge irrigation to feed growing numbers.

Bangladesh in turn suffers from flooding in the monsoon and drought and salinity ingress during the lean season, and seeks a remedy for this adverse alternation. Barring drainage from the Chittagong hill tracts, all its rivers originate in India or further up in Nepal, Bhutan, and Tibet, a factor that compels cooperation with these upper riparians for redressal of its problems.

The absolute availability of water in the GBM basin is not insufficient for the divergent needs of all the riparian states, but its seasonal and spatial distribution are skewed. Prior appropriations, mostly by India as an early starter and, in respect of Bangladesh, an upper riparian, pose problems of equitable utilization. Where water resource development has cross-country implications, issues of cost-benefit sharing would also need resolution.

From zero-sum to win-win

The clutch of disputes spawned by divergent and piecemeal approaches to these issues over the past decades hampered progress, giving rise to lingering doubts and suspicions on all sides. Bhutan was the happy exception, and commenced harnessing its water resources in cooperation with India. Elsewhere, reducing water resource development to a narrowly focused zero-sum game entailed deadlock.

After many wasted years, the impasse has begun to yield. A changing political environment with new governments in place in Nepal and Bangladesh and the initiation of the Gujral doctrine, named after the then Indian United Front foreign minister who subsequently became prime minister, facilitated the process. Inder Kumar Gujral felt India should not demand strict reciprocity from asymmetrical partners in the interest of equity and confidence-building. It should also be prepared to accept the pace and pattern of cooperation dictated by its smaller neighbours.

This approach helped move India's long-stalled negotiations with Bangladesh and Nepal from the technical to the political plane, where they rightly belong. A new willingness to accommodate the political compulsions of its smaller partners facilitated the construction of a broad framework of mutual cooperation. The Mahakali treaty and Ganges sharing treaty were respectively concluded with Nepal and Bangladesh in 1996. Both are landmark agreements and signal progress towards win-win, positive-sum outcomes. The earlier history has been well rehearsed (Verghese 1990, 1996; Verghese and Lyer 1993, 1994). Opportunity beckons, and it is now for all concerned to grasp the future.

Both the Ganges and Mahakali accords essentially deal with a single major problem, namely that of lean-season sharing of Ganges flows by India and Bangladesh at Farakka, the apex of the Ganges delta below which it fans out and enters Bangladesh, and the proposed Pancheswar dam on the Mahakali river along the Indo-Nepal border. However, the import of these agreements is far wider. They enunciate principles that should govern further cooperation on the basis of optimized, integrated basin development aimed at maximizing mutual benefits with least harm to either party.

Mahakali agreement

The Mahakali agreement, together with the letters exchanged in relation thereto, provide for a joint Indo-Nepal project on the basis of a 50:50 cost-benefit split, as the dam will straddle the border which lies along the median point of the river. Two power stations are projected, one on either bank, with an overall installed peaking capacity of between 5,500 and 6,480 MW at 20 per cent load factor operating four hours a day. These parameters will depend on the unit sizes that can be transported to the site, available cavern sizes to house the turbines, and the siting and height and, hence, reservoir capability of the proposed re-regulating dam.

A re-regulating dam could be built either at Poornagiri (180 metres high, 1,000 MW) or further upstream at Rupali Gad (about 60 metres high, 240 MW) to hold the waters passing through the Pancheswar turbines and provide regulated lean-season releases to irrigate designated commands in Nepal and India. The "total maximum net benefit" norm would favour Poornagiri over Rupali Gad, though it would cost considerably more. Both sites have to be properly investigated, and the prospects for overall funding for the total complex will have to be weighed before a final decision is taken.

Both parties are guaranteed an "equal entitlement" in the utilization of the waters of the Mahakali river "without prejudice to their respective existing consumptive uses". The Pancheswar dam will also provide a modest flood cushion.

The principles outlined commit both sides to design and operate the project as a single, integrated scheme to yield "the maximum total net benefit", with costs borne by both parties "in proportion to the benefits accruing to them". Thus, the net power benefit is to be assessed on the savings in cost compared with "relevant alternatives available"; that from irrigation on the basis of "incremental and additional benefits due to augmentation of river flow"; and that from flood moderation in proportion to "the value of works saved and damage avoided".

The avoided-cost principle of power tariff setting is straightforward enough. Thermal replacement projects could theoretically be coal, gas/oil, or nuclear fuelled, or even a hydro-electric scheme. The upper Mahakali valley above Pancheswar in India itself has a power potential of 5,000 MW or more. One norm that has been suggested by Nepalese spokesmen might be to average the sum of these comparative costs and split the savings equally (Jha 1996). According to a Nepalese estimate, Pancheswar power may cost 3.6 US cents per unit as against a replacement cost of about 8 to 8.5 cents per unit, making a difference of 5 cents.

Over and beyond cost sharing, Nepal's water requirements are to be accorded primacy in the utilization of Mahakali waters. This is a commitment that India has more-or-less accepted across the board in respect of all of Nepal's major rivers. As it happens, the total irrigable area in the kingdom is limited, barely 1.3 million hectares of this lying in the terai or piedmont before the rivers enter the Indian plain. Thus Nepal's consumptive uses will necessarily be limited. Part of the irrigation return flows will also be regenerated downstream.

The Mahakali treaty subsumes all other Indo-Nepalese agreements relating to downstream projects on the river, including the Sarada barrage (1920) and the Tanakpur barrage with a 120 MW hydro-electric capability (1995). Both these are located downstream of Pancheswar before the river finally enters Indian territory, where it is known as the Sarada.

Under the treaty, the Pancheswar project and any others that follow (such as the proposed re-regulating dam powerhouse) shall be executed and operated by joint entities established by the two parties. The treaty as a whole is to be monitored by a Mahakali Commission on the basis of "equality, mutual benefit, and no harm to either party". In case of any unresolved dispute, there is provision for binding arbitration, with the chairperson of the three-member tribunal being named by the Secretary-General of the Permanent Court of Arbitration at The Hague if necessary.

The Pancheswar project as such is to be implemented by a joint Pancheswar Development Authority to be set up under an agreed time schedule from the date of the treaty's entry into force. There has, however, been some slippage in preparation of the detailed project report because of differences with regard to the meaning of "existing consumptive uses" which are to be protected and the phasing of the power component.

Girjapur controversy

This is a misunderstanding which it is being sought to overcome. The treaty provides for sharing of the tentatively estimated Rs 12,000 crores (US$1.00 = Rs 44; 1 crore = 10,000,000) cost of the project (at 1996 prices) by both sides in proportion to the benefits received. In operational terms, the capital cost of the dam will have to be apportioned as between power, irrigation, and flood moderation. On a preliminary reckoning, the ratios may work out as approximately 80 per cent for power and 20 per cent for irrigation, barring an incidental 1 per cent for flood moderation. On this basis, the burden of the irrigation component of 20 per cent of the total cost should be shared 50:50 provided the water benefit is utilized in equal proportions. However, should Nepal as the upper riparian not fully use its due share of the augmented flows, the side-letters exchanged "preclude the claim in any form by either party on the unutilised portion of the shares of the Mahakali river of that party".

Nevertheless, should India consequently use more than its agreed half-share of the augmented lean flows of the Mahakali below Pancheswar, it will be liable to bear an equivalent proportion of the capital cost attributable to the irrigation component of the dam. This is more-or-less established international practice and is perhaps not really in dispute.

Nepal is firm in stating that the ambit of the treaty is limited to that part of the river that more-or-less defines the Indo-Nepal border, which stretch is known as the Mahakali. It does not extend to the upper tributaries or the lower reach of the river, which is called the Sarada after it leaves Nepal and flows exclusively through India until it falls into the Ghaghra. Nepal thus protects the stipulated and existing quantum of irrigation from the Tanakpur and Banbassa barrages, and the upper Sarada canal taking off from Banbassa and commanding an area of 1.6 million hectares. Both these are specifically mentioned in the agreement. By the same token, the argument proceeds, it is not committed to protecting any "existing uses" from the lower Sarada or Girjapur barrage 160 km below the Nepalese border, which commands an additional 2 million hectares.

The Girjapur barrage is fed by a diversion from the Karnali, a far larger river, through the Sarada-Sahayak canal for just over eight months in the year. But on account of the heavy silt the Karnali carries between June and September, the Girjapur barrage draws Sarada waters during this period. Assuming 75 per cent dependability and the Pancheswar dam in place, simulation studies indicate that there should normally be sufficient water to meet the requirements of the Girjapur barrage together with regeneration and other inflows into the river below Pancheswar. There could, however, be occasional distress in July when paddy is sown, and maybe in October when the crop matures. The water passing through the Pancheswar turbines during these months may not by itself suffice to meet existing irrigation uses in the Tanakpur, Banbassa, and Girjapur commands in all years, coinciding as they largely do with the filling period of the reservoir. The problem, if any, is therefore related to the operation of the dam as a peaking station rather than to overall water availability.

It is perhaps for this reason that India had proposed phasing the power component in two stages, with an installed capacity of 2,000 MW in the first instance and 5,500 to 6,480 MW in a second stage with a re-regulating dam in place. Nepal, on the other hand, would prefer to maximize total net (power) benefit, with Pancheswar as a peaking station from the very inception with a re-regulating dam to take care of irrigation commitments. India also appears to be coming round to the view that this would indeed be logical, as its power requirements are both urgent and important, while irrigation need not suffer as the re-regulating dam can be investigated and constructed before Pancheswar is commissioned.

Further, since Nepal's own planned Mahakali command of 93,000 hectares will take some years to develop fully, especially at the very high irrigation intensity proposed, any problem of overall water availability in the near future is likely to be further minimized. This offers added lead time for detailed studies and finding suitable alternatives to the extent required.

Water rights, not ownership

Any notion, whether implicit or explicit, that the Mahakali treaty implies 50:50 ownership by the two sides of the natural river flows of the Mahakali is mistaken. Ownership of water by the upper riparian was a claim embodied in the Harmon doctrine of territorial sovereignty asserted by the United States in 1898 in the course of a dispute with Mexico over the waters of the Rio Grande. The doctrine was subsequently given up. Though asserted from time to time, it has never found acceptance anywhere in the world as a legal principle any more than its mirror opposite, namely the theory of territorial integrity advocated by lower riparians to claim a prescriptive right to pristine stream flows entering their territories.

Section 3(a) of the side-letters exchanged with the Mahakali treaty leaves no room for ambiguity. It explicitly states that the "irrigation benefit shall be assessed on the basis of incremental and additional benefits due to augmentation of river flow". Hence any reference to ownership of half-shares relates only to augmented flows as a result of the storage created behind the dam, and not to the natural flow of the river. This same principle is set out in the understanding reached by the joint team of experts on the Sapta Kosi project at a meeting held in Kathmandu in January 1997.

The live storage to be created in the Pancheswar reservoir is estimated to yield a lean-season flow of around 582 cumecs. There is no problem in dividing the augmented flow equally between the two sides. However, Clause 3 of the treaty states that "both parties agree that they have equal entitlement in the utilisation of the waters of the Mahakali river". Nepal accordingly claims an equal share in the natural flow of 144 cumecs generated in the catchment between Pancheswar and Banbassa as well – or, in other words, in the overall lean-season availability of 726 cumecs at Banbassa.

India requires 448 cumecs at Banbassa for the upper Sarada canal command (of which 200 cumecs is to secure a higher intensity of irrigation). Nepal demands roughly 160 cumecs to meet existing and planned uses. Together with the 10 cumecs mandated by the treaty to maintain the ecology of the Mahakali, this adds up to 618 cumecs. This still leaves a "surplus" of 108 cumecs, as against the requirement at Girjapur of 228 cumecs. The deficit of 120 cumecs should normally be made good by additional flows from the free catchment below Banbassa and through regeneration, except in lean years during June and possibly October as explained earlier.

Water availability to meet all existing uses is therefore more-or-less assured. What is at issue is the sharing of the capital cost of the Pancheswar project in relation to its estimated water component of approximately Rs 2,400 crores on the basis of the assessed irrigation benefits to the two sides. If the base figure of water availability is taken as 726 cumecs, India's actual utilization may work out to around 75 per cent, rather than the stipulated half-share, requiring it to pay some Rs 1,800 crores towards the capital cost of the irrigation component of the project.

What is at issue is this enhanced capital cost liability – which, however, will be less than the opportunity cost of a year's delay should the impasse be prolonged. Further, it would be counterproductive to disregard the political compulsions that weighed with the Nepalese government in projecting the interpretation of the treaty that it did in order to secure its ratification by the constitutionally required two-thirds majority.

It would be best to accept the situation as it has developed, while making it clear that it does not constitute a binding precedent. Both sides should be wiser for the experience. Certain basic principles must be reiterated and asserted for the future: water rights should not be confused with ownership of flowing water. Thus, Nepal cannot use the Mahakali "precedent" to claim ownership of the natural flows of the Kosi, Gandak, or other rivers. To accept this principle would be revert to the Harmon doctrine and could totally undermine India's international and federal water relations, with disastrous consequences.

Secondly, India must accept that prior appropriation is not an absolute principle if it results in what might approach expropriation in relation to late developers. Reasonable adjustments would be necessary should any such contingency arise in order to ensure equitable apportionment.

Thirdly, there could be an arguable case for payment of a one-time "royalty" for the dam site, a scarce natural resource and wasting asset to the extent that it tends to silt up over time. The Mahakali treaty is for 75 years: this may be taken as the life of the Pancheswar site over which period the dam might be depreciated, as against 50 years in respect of many Indian projects.

The power scenario

While the treaty bars any claim by Nepal for its share of unutilized augmented flows, it may, if it so desires, sell any or all of its share of power under the terms of the 1996 Indo-Nepal power trade agreement. This permits the two governments, semi-official agencies, or private parties to enter into agreements for the investigation, construction, and generation of power and its transmission between the two countries or to a third country.

Any authority or corporate entity in Nepal may accordingly use or promote inter-country grid links to export energy to consumers in India, or to wheel power through India to Bangladesh or Pakistan. Such possibilities are not far-fetched by any means. While the Indian market is going to remain power hungry in the foreseeable future, it would nevertheless provide assurance to Nepal were it able to sell power to third countries, breaking a potential Indian stranglehold as a monopoly buyer and widening and deepening its economic relations with other SAARC members.

Nepal has a hydro-electric potential of 83,000 MW, of which some 42,000 MW has been assessed as being techno-economically feasible. India currently has an installed generation capacity of around 90,000 MW with a significant energy and peaking deficit. According to earlier forecasts, the country must add another 112,000 MW by 2007 to match the expected growth of the economy. Not all of this will come from hydropower, but there is widespread agreement on enhancing this component to arrest and reverse what has been a declining hydro-thermal mix.

Much of this additional capacity will come from within India itself, as large hydro and pumped-storage potential remains unexploited. Nevertheless, what this suggests is that Nepal and Bhutan's estimated hydro potentials could theoretically be absorbed in their entirety by the Indian market within the next few decades. This is a market opportunity that neither country nor international power companies and equipment suppliers are likely to want to miss, within the limits of economic, environmental, and political prudence.

Power development in Bhutan

Bhutan's hydropower development is, meanwhile, proceeding apace with Indian technical and financial assistance. The 336 MW run-of-the-river Chukha project on the Wangchu (Raidek) was commissioned in 1987 and has transformed the Bhutanese economic outlook in terms of energy, income, exports, technical development, and forest conservation. Whatever Bhutan cannot absorb is fed into the Indian grid. Encouraged by this success, the Wangchu cascade has been investigated and a series of projects are to be implemented in stages.

The Tala project (1,020 MW) is under construction, with the downstream Chukha Stage III (900 MW) and Bunakha (180 MW), in the head reaches of the river, to follow. A smaller project on the Kurichu (45 MW) is also under way in eastern Bhutan, while a project report with regard to the larger Sunkosh project (4,060 MW) is in an advanced stage of preparation. This adds up to a tidy 6,500 MW which in theory could be on stream by 2015–2020.

It would, however, be simplistic to assume that whatever is proposed will necessarily be approved and implemented without difficulty. The king of Bhutan is not anxious to catapult his unspoilt country into instant modernization, mindful of the social and cultural trauma that he fears this might entail. He would prefer to hasten slowly, ensure sustainability, and seek an outcome that assures high "gross national happiness" rather than a larger gross national product.

Nepal and India, too, will encounter environmental and other objections as they proceed with exploiting their water resources.

Ganges treaty

The Ganges treaty, likewise, has opened new windows of opportunity. While the principal focus is on "a fair and just solution" with regard to sharing the lean flows of the Ganges at Farakka "in a spirit of mutual accommodation", this is not the sole concern. It envisages the sharing of all the other international rivers flowing from India to Bangladesh. These include the Brahmaputra, the Barak/Meghna, and the Teesta, as well as 54 other medium or small rivers which have sometimes been grouped in clusters.

The call for "mutual accommodation" in regard to the Ganges again suggests a political rather than a technical approach to the problem. Further, the treaty enjoins "optimum utilisation" of the water resources of the "region" through integrated flood management, irrigation, "river basin development", and generation of hydropower for mutual benefit. It underlines the principles of "equity, fairness and no harm to either party" and recognizes the need on the part of the two sides to cooperate in finding a solution "to the long-term problem of augmenting the flows of the Ganga/Ganges during the dry season".

Prior to this accord, Bangladesh had felt politically unable meaningfully to discuss and conclude agreements with India pertaining to any river other than the Ganges, let alone on matters unrelated to water resources such as transit and wider economic cooperation, until the Farakka issue was first settled or clearly on the way to settlement. This brought about an almost complete stalemate in Indo-Bangladesh relations but for some progress on trade under the rubric of SAARC. The Ganges treaty has now unlocked all doors, though certain internal political impediments may remain to be overcome on either side.

Unlike the 1977 accord, the 1996 treaty is for 30 years and separates sharing from augmentation. Further, it indicates flows in terms of 50 per cent rather than 75 per cent dependability as heretofore. This criterion enhances the yield below Farakka, as the assurance level has been reduced from three to two years out of four. The purpose was presumably to facilitate the notional allocation of a minimum 35,000 cusecs to Bangladesh during March-April, in order to match and even top the minimum 34,500 guaranteed to it under the 1977 schedule, while giving India the 40,000 cusecs it claims to safeguard Calcutta port. There is also a suggestion that the Gorai, a major deltaic spill channel serving the southwest region, can only draw water when the Ganges discharge exceeds 30,000 cusecs.

The actual schedule of distribution is set out in Annexure I to the Treaty (see Appendix). This stipulates an allocation of 40,000 cusecs to India and the balance to Bangladesh when arrivals at Farakka are in excess of 75,000 cusecs; 35,000 cusecs to Bangladesh and the balance to India when flows are between 70,000 and 75,000 cusecs; and equal sharing when flows are 70,000 cusecs or less, subject to the condition that each side shall be guaranteed 35,000 cusecs in six alternate 10-day periods between 1 March and 10 May.

The implications of this formula are illustratively set out in an "indicative schedule" in Annexure II, which mentions certain numbers "if actual availability corresponds to average flows of the period 1949 to 1988" (see Appendix). For its part, India is required to make "every effort" to protect existing flows at Farakka.

The treaty provides that should flows at any time fall below 50,000 cusecs, "the two Governments will enter into immediate consultations to make adjustments on an emergency basis in accordance with the principles of equity, fair play and no harm to either party". Both sides are pledged to cooperate in finding a solution to the long-term problem of augmenting the dry flows of the Ganges.

Three problems

The working of the agreement was soon to reveal certain shortcomings that had been glossed over. First, while the treaty is clearly based on the sharing ratios set out in Annexure I, the more iffy, politically designed Annexure II was projected as if it contained the essence of the treaty. When the "illustrative flows" failed to match the reality over two 10-day periods in 1997 owing to a natural hydrological deficit, with flows briefly dipping marginally below 50,000 cusecs, there were accusations of default by India and bad faith on its part in abstracting more water upstream.

The treaty specifies "immediate consultation" in the event of flows dipping below 50,000 cusecs, but does not prescribe any particular mechanism or formula such as the 80 per cent guarantee clause embodied in the 1977 accord. The fall of the then Indian government in February–March 1997, precisely when flows did dip below the emergency level, and the absence from Delhi of the key Indian technical adviser, who was busy attending the UN session convened to adopt the draft declaration on the non-navigational uses of water, did cause an unfortunate hiatus for a few days. Fortunately, the flows soon rose above 50,000 cusecs, thereby solving the problem.

The switching of 10-day flows to guarantee one side or the other 35,000 cusecs during six alternating 10-day periods is technically unsafe, as the walls of the Farakka feeder canal could collapse with such abrupt changes in levels. This was known to both sides from previous negotiations. Even so, a stepped pattern was written into the treaty instead of a smoother gradient spread over a few days whereby the quantum of water delivered would have remained the same though with a different 10-day flow pattern. The Indian side had perforce to stagger the reduction and augmentation of releases over some days. This altered the pattern of scheduled deliveries in a few 10-day periods, with Bangladesh being compensated for any scheduled deficit in supplies in the ensuing period. This was cited as a treaty violation, though in point of fact the total quantum of water released to Bangladesh over the lean season as a whole was a little more than stipulated in Annexure II.

While releases and arrivals were being jointly monitored at Farakka and Hardinge Bridge respectively, a third problem arose as the character of the river changed from affluent to effluent in the course of the lean season. Thus, while there was regeneration up to the end of February between the two observation points, with Hardinge Bridge recording higher figures than Farakka, there was degeneration for some time thereafter, with Hardinge Bridge recording 20 to 40 per cent less than the release figure at Farakka. There was an outcry that India was manipulating releases at Farakka despite joint monitoring.

Since there was no observed diversion of flows on either side between Farakka and Hardinge Bridge, some 120 km apart, the inference is that the "missing cusecs" may have disappeared into aquifers that began to be emptied after late February with groundwater pumping to irrigate the boro crop.

Indeed, such a possibility was forecast by the influential Bangladesh daily, *Sangbad*, on 2 January 1997. The paper reported that:

> the Government [of Bangladesh] should pay special attention to the underground water level in the northwestern region for proper maintenance and usage of water flow received after signing the Ganges Water Agreement. Hydrologists have expressed their concern that during the lean period in March–April the exact quantum of water released from Farakka under the newly signed agreement may not be received at the Hardinge Point. This may happen as the affluent river condition which was prevailing five years ago is not there; rather influent river condition is existing ... [*sic*]. Opposition parties may misinterpret this to put the Government into trouble.

The paper explained that over 100,000 tubewells had been drilled under the Borendra project in a 7,692 square kilometre block from opposite Farakka to Hardinge Point. As much as 31,700 cusecs of water were being extracted here, against a safe limit of half that figure as advised by technical experts. As a result, the groundwater table had fallen from 25 feet to 60 feet over the past five years. The evidence was not conclusive, and a joint investigation has been mounted and will establish how and on which side there may have been losses to aquifers.

Joint consultations for amending the stepped release schedule over the six prescribed alternating 10-day periods, and for instituting a self-actuating mechanism and formula should flows dip below 50,000 cusecs, did not take place during the 1997 flush season despite a meeting proposed by India. However, a greater degree of transparency has been introduced in the joint observation procedures, with Bangladesh officials permitted to visit the observation sites at any time of day or night. It is now open under Article X of the treaty for either party "to seek the first review after two years (December 1998) to assess the impact and working of the sharing arrangements". Fortunately, arrivals at Farakka up to the middle of March 1998 were more than satisfactory.

It must also be said that both governments have throughout staunchly defended the treaty despite virulent critisism by political opponents.

The Gorai hump

There is a further and, in a sense, more complex problem. Bangladesh's grievance about diversions by India from the Ganges at Farakka has revolved around the acute distress said to have been caused in the south-west Khulna region on account of salinity ingress and a shortage of water for agriculture, fisheries, navigation, and sustenance of the sundari mangrove species. This area, because of the Gorai spill, which delivers upland fresh water supplies to the region, is left high and dry as the Ganges recedes.

While this is so, it would be erroneous to attribute the problem exclusively or mainly to diversions at Farakka. The entire Ganges system has been shifting east and north as a secular trend over the past century and more. The Bhagirathi, the westernmost spill, was the first casualty. Other streams, moving further east, have progressively deteriorated as the Ganges has shifted course. This has resulted in the closure of the Gorai mouth by a hard five-metre silt plug that is now said to measure almost 30 km from the outfall point to Gorai bridge.

The Bangladesh Land and Water Resources Study (Ministry of Water Resources 1972) is instructive. Para 6.12 has this to say:

> Present dry season flows in the Gorai River are often decreased by sandbars that develop at the Ganges offtake. Without [the] sandbar, the minimum discharge varies historically between 10,000 and 18,000 cfs [cusecs], but with [the] sandbar the discharge had decreased to 5 cfs in February–May 1950, 2,900 cfs in January 1952, and to 500 cfs in May 1954. However, the effect of sandbars could be eliminated by dredging and a minimum discharge of some 15,000 cfs could be assured under the present circumstances.

Though Farakka diversions commenced in 1975, agreements with India were in place guaranteeing far more than 15,000 cusecs of Ganges flows below Farakka between 1977 and 1988. Nevertheless the Gorai deteriorated and has in recent years stopped drawing water even in January, when the discharge ranges between 50,000 and 60,000 cusecs. This being so, only a limited quantum of Ganges treaty releases during the lean-season trough enter the Gorai and none at all below a certain threshold related to the hump at the outfall, the elaborate formulations of alternating 10-day releases of 35,000 cusecs to either side in March–April notwithstanding.

In view of this situation, East Pakistan/Bangladesh in 1963, 1964, 1968, 1984, and 1994 had surveys and investigations carried out by local and foreign consultants to construct a Ganges barrage to pond the river and force water into the derelict Gorai channel. This would variously irrigate

670,000 to 1.5 million hectares of land in the south-west region and resuscitate the Ganges-Kobadak project. The proposal has come alive again with the signing of the Ganges treaty.

Bangladesh has formally approached various multilateral agencies and India for assistance in funding and carrying out a feasibility study on the Ganges barrage project and its possible environmental impacts. The response encouraged the government to convene an international conference in Dhaka in March 1998 to pursue the matter. India has formally conveyed its willingness to share its design and technical experience regarding Farakka with Bangladesh.

A Gorai restoration programme is already contemplated, with World Bank and Dutch assistance. This envisages capital dredging of the Gorai outfall to permit lean-season inflows from the receding Ganges, followed by maintenance dredging to maintain this channel. This is seen as perhaps no more than an interim solution pending ponding the Ganges behind a barrage at Pangsa, 10 km below the Gorai outfall, thereby enabling the backwaters to flow into the moribund Gorai. However, instead of constructing an elaborate and expensive canal network, it is proposed that the Gorai restoration project should clear the clogged outfalls of all lower spill channels to restore as far as can be the original ecology of the Gorai system. This would enable farmers to irrigate their fields by groundwater and river lift. Such participative development is seen as more likely to ensure maintenance of the environment.

In view of these changed parameters, the World Bank, ADB, and Japanese government have persuaded Bangladesh to undertake certain preliminary studies through the Water Research and Planning Organization (WARPO) before they are willing to fund an actual Ganges barrage feasibility study. This proposal has been accepted, though it could mean a possible delay of up to a year or more before the commencement of that study.

A more balanced perception is now beginning to emerge in both Bangladesh and India, some hawks apart. This augurs well for future cooperation. While it is necessary to understand and draw lessons from the past, it is more important to seize the opportunity and chart the way forward.

Critics on both sides may chafe at one country or the other having received what they perceive to be a raw deal. The interests of Bihar and Uttar Pradesh have not been prejudiced. On the other hand, some in Bangladesh believe that sharing the residual flows of the Ganges at Farakka, after generous abstractions in upper India, is *per se* unfair. Is this really so? The Ganges has a smaller annual discharge than the Brahmaputra, which, virtually untapped as it is throughout its course even today, drops to not less than 110,000 cusecs at the Indo-Bangladesh border when it troughs in early February. If one makes the somewhat generous assumption that the Ganges' lowest virgin flow at Farakka in March–April is as much, a 25–30 per cent share of this quantum is going to Bangladesh, which has less than 5 per cent of the basin area and 10 per cent of its population as compared with India. Is this unreasonable? This calculation obviously ignores the groundwater factor, but nevertheless offers an admittedly crude basis for making a comparison.

New role for Calcutta port

There are some in India who fear that Calcutta port will choke for lack of minimal headwater supplies during the critical lean months. While diversions into the Bhagirathi-Hooghly at Farakka have reduced salinity and enhanced drafts at Calcutta, the limits of significant improvement have been reached. Moreover, Calcutta port has been overtaken by events. The nature of shipping and freight has changed. Bends and bars in the river preclude vessels beyond a certain size negotiating the fairway to Calcutta. Haldia, 120 km downriver, has in consequence outgrown its auxiliary role to become the major partner in the twin-dock Calcutta port system. Vessels sailing to or from Calcutta have to be lightened or topped up here or elsewhere.

Freight composition has also changed. Bulk cargoes such as ore and coal move through Haldia, Paradeep, and other ports; liquid cargo is piped ashore at Haldia; and food grain imports have been replaced by grain stocks from Punjab, Haryana, and western Uttar Pradesh. Moreover, Calcutta is today no more the prime, let alone sole, ocean port at the head of the Bay of Bengal. It could, however, become a vibrant inland waterport handling a far larger volume of containerized and other traffic than at present. The city, the run-down dynamo of eastern India, cannot be restored to health if it continues to be miscast in the role of a great ocean terminus that it is simply unable and not called upon to play, as there are alternative ports and options. The pretence that more flushing will somehow improve matters and that reduced headwater flows for two months during the critical lean season will spell disaster for Calcutta are myths. The opportunity cost of flushing water is very high and rising fast. Bengal farmers are already pumping up to 8,000 cusecs from the Farakka feeder and resuscitated Bhagirati to water their fields. The Hooghly faces draft and shoaling problems well below Calcutta, and even at Haldia, and dredging costs have mounted.

India desperately needs more port capacity countrywide, and certainly so at the head of the Bay of Bengal. Calcutta, and even Haldia, cannot cope. Not only must Haldia be greatly expanded, but there is a clear need for more deep-water port capacity in the vicinity. Several proposals are under consideration, including a small container terminal opposite Haldia at Kulpi, a major port on Sagar island at the mouth of the Hooghly, and perhaps a single-buoy mooring further out to sea. Against a mere 23 million tonnes handled by Calcutta-Haldia in 1997, freight forecasts indicate an overall bay port capacity requirement of 50–100 million tonnes and more in another decade to 25 years.

In such a scenario, a refurbished polynodal inland-water Calcutta port strung over the reach from Haldia to Farakka could hum with activity as a feeder and distributive node with containerized and other barge trains plying extensive inland, estuarine, and coastal waters. These vessels could negotiate the Farakka lock and travel up the Ganges to Patna and Allahabad, or go through the Sunderbans and Bangladesh up the Brahmaputra and Meghna/Barak to Guwahati-Dibrugarh and Karimganj, or move down the coastal canal to Orissa and further south along the Buckingham canal, both derelict systems that are being or could be revived. The existing rail-road transport corridors are under heavy pressure, and will need large investments and land for modernization and expansion. Some of that investment could usefully be diverted to developing waterways which are fuel-efficient and environmentally friendly.

Augmentation possibilities

The Ganges treaty speaks of augmenting lean-season flows below Farakka. Conservation and demand management on both sides of the border would be one route to this end. But additional flows could emanate from storages and diversions. The Sunkosh hydro-electric project (4,000 MW) investigated in Bhutan suggests the possibility of utilizing the existing Teesta-Mahananda barrages and canals as a ready-made alignment to divert up to 12,000 cusecs to Farakka. One advantage would be the prospect of utilizing some of this to firm up supplies for the twin Teesta projects in north Bengal and Bangladesh, which presently face significant shortages. A longer feeder canal realigned south of certain major game sanctuaries would save them from harm, but would still entail slicing through several tea plantations, whose owners would have to be suitably compensated. Environmental opposition and Bhutanese political sensitivities will have to be overcome.

Some of the proposed high dams in Nepal, especially that on the Sapta Kosi, could also augment flows below Farakka despite utilization in India. It will be some time before these projects are commissioned. There is no reason why Bangladesh's interests should not be accommodated, with an appropriate tripartite sharing of costs and benefits. It would be fanciful to suppose that all such projects will be designed exclusively to serve the purpose of augmentation below Farakka. It is hydropower that makes them viable, with regulated releases of water for irrigation and flood moderation as by-products. India's energy market and, to a lesser extent, its still unsatiated irrigation needs are therefore likely to remain the determining factor in both the project design and benefit mix after fully meeting Nepal's own requirements.

Obviously, Bangladesh's water requirements are not limited to the Ganges alone. Of a cultivable area of just under 9 million hectares countrywide, only 3.14 million hectares lie in the Ganges basin as against 3.75 and 2.61 million hectares in the Brahmaputra and Meghna basins respectively. The Chittagong hill tracts and the plains district stand apart. However, almost the entire GBM flows enter Bangladesh from India or beyond. Regional cooperation is therefore a compelling necessity, whether for managing floods or drought.

There was for a while a feeling in Bangladesh that it could perhaps manage on its own with the flood action plan. This concept was unrealistic, and has now been abandoned in the manner originally conceived. The sharing of the lean flows of the Teesta and possible joint operation of the twin Teesta barrages are now being explored, alongside prospects of augmentation from the Sunkosh or any marginal Teesta storages possible in the Darjeeling-Sikkim hills.

Brahmaputra-Barak options

The other possibility would be a barrage across the Brahmaputra at Jogighopa in western Assam, with a canal that could divert water into the Teesta and Mahananda, or down the Karatoya-Atrai. This would augment lean flows in these rivers and, through the latter, in the Ganges. This is not to revive the hypothesized 100,000-cusec Brahamputra-Ganges link canal proposed in 1978, which India has withdrawn, but to explore lesser variants along several different alignments as may be jointly evolved with Bangladesh. A diversion canal from a Jogighopa barrage near Dhubri in Assam could lead to Bangladesh's proposed Ganges barrage. It could be aligned to follow or join the Karatoya-Atrai or other river courses, and would irrigate considerable areas and stimulate fisheries and navigation en route.

A Brahmaputra barrage at Jogighopa could be a less expensive proposition, for reasons of topography, than a possible structure lower down. This might interest Bangladesh, as it has at various times contemplated a barrage at Bahadurabad from which it could irrigate the central part of the country and service its Ganges-dependent region as well.

The Brahmaputra accounts for 29 per cent of the total runoff of all of India's rivers, representing a huge per capita or per-unit-of-arable-land water availability that will always remain surplus to any possible consumptive uses within its own international basin. Hence India's long-term interest in developing a national water grid and diverting Brahmaputra waters west and south as part of this perspective to meet emerging water shortages in large parts of the country.

The notion that the Brahmaputra has surplus water is, however, challenged in Dhaka. In 1987–88 Bangladesh informed India that it would be satisfied with a 25,000–27,500 cusec minimum lean-season share in the Ganges provided it was given a half-share in all the other rivers and a 75 per cent share in the Brahmaputra. Its formula for the Brahmaputra was that each country might be allocated 25 per cent of the lean flows for consumptive uses, with the remaining 50 per cent being left to find its way to the sea as an ecological necessity. This thesis is said to find support in a study by a firm of international consultants which has reportedly cautioned against any depletion of the existing virgin lean-season flows of the Brahmaputra lest this harm the environment and permit salinity ingress.

Prime facie this would seem altogether too sweeping a claim, and requires detailed justification. Nevertheless, India's ability to divert Brahmaputra waters westwards is limited, while any proposed storages built in Arunachal or Bhutan for purposes of hydro development, flood management, and navigation would greatly augment lean-season flows.

The original Indian proposal for two mega-dams on the Dihang, the main stem of the Brahmaputra, and the Subansiri have had to be dropped because of strong opposition in Arunachal, as five politically important townships would have been submerged. The proposals are now being reworked in cascades of three dams each. These would avoid submerging any township while more-or-less maintaining a significant part of the originally proposed storage and installed hydro-electric capacity (around 20,000 MW). Flood peaks would still be lowered, while regulated flows would improve navigability in the Brahmaputra below Dibrugarh. Investigations are in progress, and the first feasibility and environmental impact studies should be available by 2000.

The Barak, as the Meghna is known in India, is also proposed to be harnessed at Tipaimukh in Manipur, very near its trijunction with Mizoram and Assam. The project is expected to provide a substantial flood cushion and an installed capacity of 1,500 MW, with added irrigation and navigation benefits which could be extended to Bangladesh, though no cost-benefit studies have been made in this regard. Such a study of downstream impacts has now been informally proposed to Bangladesh.

Unfortunately, the Tipaimukh project has run into difficulties in Manipur on issues of submergence and displacement, which would partly appear to stem from misperceptions caused by lack of transparency at earlier stages. This underlines the importance of taking the local people and other stakeholders into confidence from the outset.

Strategies for Nepal

In Nepal, as in Bangladesh, there is every reason for expedition in harnessing the eastern Himalayan rivers. The Mahakali treaty is a framework agreement which sets out general parameters for overall water resource cooperation. The present impasse over Pancheswar is no more than a hiccup. Political uncertainties have clouded negotiations, and all three countries need to move forward resolutely, riding the momentum generated by the 1996 agreements. Both Nepal and India should use Pancheswar to make a new beginning rather than continue with tired arguments. The joint DPR (detailed project report) must be expedited in a spirit of mutual confidence. Overdependence on foreign consultants seeking a stake in the action can prove counterproductive. Hydropower is to Nepal what oil is to the Gulf. Possession is not enough; on the other hand, its exploitation could be transforming. The opportunity cost of delay can therefore be high.

The Indo-Nepal power trade agreement opens the door to private investment in water resource development. Private entrepreneurs are now allowed to sell energy to third parties. India need therefore no longer be a monopoly buyer. Bangladesh and Pakistan could also become partners with Bhutan and Nepal in a subcontinental power grid that would impart great flexibility and reliability to the system.

Apart from Pancheswar, there is an Indo-Nepal agreement for joint preparation of a detailed project report on the Sapta Kosi high dam and Sun Kosi hydro diversion into the Kamala basin which is to be financed by India. The inception report sets out certain broad parameters, including investigation of the feasibility of a navigation link by river or canal from the dam to the Ganges to give Nepal an outlet to the sea. However, progress has stalled with the impasse over the Mahakali. Since the Sapta Kosi project offers some real assurance of augmentation of Ganges flows below Farakka, Bangladesh should be brought into the discussions at an appropriate stage so that its interests can be built into the overall design in accordance with agreed cost-benefit principles.

With Nepal throwing open water resource development to the private sector, two hydro-electric projects, Khimti (60 MW) and upper Bhote Kosi (36 MW), have been taken up. An American corporation has staked an interest in the giant Karnali project (10,800 MW), which may only be constructed in stages. Upper Karnali (300 MW) might get off the ground sooner.

Other private parties are reportedly interested in the 404 MW Arun-3 which was earlier canvassed by the Nepal Electricity Authority but finally rejected by the World Bank and ADB, its principal sponsors, on economic considerations. The long access road to the site was itself estimated to take six to seven years to complete and, as a front-end investment, threatened to price the power generated out of the market unless spread over the larger Arun cascade.

The need for an immediate replacement project to meet Nepal's load demand after 2000–02 was met by taking up the 144 MW Kali Gandaki-A project. The Australian Snowy Mountain Electricity Corporation (SMEC) has come to an agreement with the Nepal government on building a 750 MW hydro project on the west Seti, a tributary of the Karnali. A 30-year build-operate-own-transfer arrangement is mooted, with a tax on power exports. The SMEC is seeking prospective buyers for the energy to be produced among north Indian state electricity boards and bulk consumers.

The Nepalese government estimates that regulated releases from the west Seti will augment lean-season flows in the Karnali by some 30 per cent, which could beneficially be used by India some time in the future. In order to obviate any such an unrequited benefit, some in Nepal would like to explore a trade-off with India. They argue that certain medium irrigation projects earlier proposed by the kingdom on the Babai and Kankai were shunned by prospective foreign donors such as the ADB and Kuwait Development Fund after being informed of possible adverse effects on existing or proposed uses in Uttar Pradesh and Bihar. Hence it is being informally suggested that possible losses to India should these projects be implemented, or from the proposed Bheri hydro project which would divert water outside the Karnali valley, be set off against the augmentation of Karnali flows from the west Seti. Such a package merits consideration. Given the gestation period involved in any of these projects, India could switch irrigation in small terai commands to groundwater lift from the *bhabbar* springline that runs along this alignment. At a pinch, Nepal and India could trade water for energy.

Some of Nepal's irrigation needs from certain smaller schemes could over a longer term be subsumed in the larger commands created by mega-projects like Sapta Kosi and Karnali. These alternatives are not mutually exclusive and could optimize the system over time.

Not unexpectedly, debate has meanwhile been joined in Nepal on the merits of embarking on any of these mega-projects. The environmental objections will be examined later. These apart, critics fear that the kingdom will become hostage to the Indian economy; that Nepal must learn to walk before it can run and should appropriately confine itself to small and medium projects in the immediate future until it is ready and able more confidently to embark on larger ventures; and that mega-projects could create social and regional distortions as annual investments and overall costs would be larger than the country's current annual budget or GDP. As a wag put it: "Will Nepal own Karnali or Karnali own Nepal?"

These fears may not be altogether unreal, but are exaggerated. Oil or gas can be left underground and exploited when required and marketed over vast distances. No so falling water. Hydro-electric energy not generated in any one season is lost for ever. Further, Nepal, like the rest of the GBM region, cannot afford to forgo the leverage of hydro development to roll back poverty and improve the quality of life of its growing population. And environmental constraints are likely to increase rather than diminish with time. As for regional imbalances, these can be evened out through appropriate policies and by investing income from mega-projects in other regions and sectors. Likewise, sale of large blocks of power to India would create a healthy relationship of interdependence that would compel greater respect for Nepal in India than if the latter's stakes in the kingdom are marginal. Building partnerships is good diplomacy.

The feeling that Nepal's earlier innocence was cunningly exploited by India in the Kosi and Gandak projects is again somewhat exaggerated. Nevertheless, that perception is a reality and can best be dispelled by the manner in which India now approaches its smaller neighbours. This is what the much-misunderstood Gujaral doctrine sets out to do by eschewing absolute reciprocity in dealing with asymmetrical relationships. That policy has paid off, and should be pursued. Seeming concessions in the short run will bring home dividends in time.

Environmental factors

The environment issue must of course be addressed. The Himalayas are a dynamic tectonic zone and, additionally, are under severe pressure over large areas from population and livestock growth, falling land-man ratios, unsound land use, cultivation of marginal slopes, adverse development impacts from careless road-building and mining, and excessive logging and felling of trees that have resulted in barren hills, erosion, degradation, the drying up of springs, and receding glaciers. All this may be true; but communities and governments are learning to do better. Some of Nepal's community forestry (and water) management programmes, like some of India's joint forest management schemes, point to new directions of participative management of common property resources. Watershed and biodiversity management are aiding conservation and ecological restoration.

All new water resource projects mandate environmental impact studies and corrective action. These range from catchment area treatment and rim stabilization to reduce siltation to compensatory afforestation and protection of endangered faunal and floral species. Upper catchment and downstream effects are also being more carefully studied and rigorously addressed. Nevertheless, there is no getting away from the fact that tectonic shaking and mass wasting of the fragile Himalayas bring down more debris and silt into the rivers by far than the sum total of all man-made interventions. Dams will silt; but the process can be slowed down. There is no reason therefore to forgo projects whose overall social and economic cost-benefit ratios, both direct and indirect, are positive. Dam safety is obviously important, and modern technology now permits construction within acceptable limits of safety and social and political tolerance.

If there is a cost of doing something, the cost of not doing is all too often overlooked. This can be far greater. Many of those who some years ago worried about the negative environmental impacts of large dams are today more concerned about the greater danger to humankind from global warming through greenhouse gas emissions from coal-fired thermal stations. Hydropower, with nuclear energy, constitutes the cleanest form of electric power. Non-conventional and newer energy sources are still far away from being affordable or practical alternatives on the scale required.

Displacement of populations by submergence behind dams is a major concern. Here, the compensation offered and the resettlement and rehabilitation package has to be generous and caring. In third-world countries like Nepal and India, the involuntary migration of Malthusian refugees from degraded lands with shrinking economic opportunities imposes untold misery on millions. Only development can arrest, even if not reverse, this sorry process, and water resource development in turn more specifically redresses the accompanying gender inequity that condemns women to a lifetime of servitude in fetching fuel, fodder, and water over ever-increasing distances.

Insistence on rehabilitating displaced populations on the land is mistaken, except where land is readily available. Land is no more than a synonym for employment and social security, and could certainly be one but not necessarily the sole avenue for rehabilitation in densely populated regions. The gestation period before the dam rises and the reservoir begins to fill is sufficiently long to provide for education, training, or re-training in marketable skills and investments that generate gainful employment in areas including the pisciculture, tourism, and watershed management nurtured by the project.

Area development opportunities

Pancheswar offers India, and certainly Nepal, an opportunity to make a new start in reconceptualizing water resource projects and rehabilitation. Large dams in remote mountain locations essentially entail area development. Even before work can commence at the dam site, a great deal of basic infrastructure has to be put in place: access roads and bridges, electricity, water supply, telephone connections, medical facilities, transport and market linkages, and so forth. The intrusive process of opening up hitherto remote and neglected areas, environmental impacts, and displacement are disruptive of cultural networks and traditional patterns of living. All this could mean trauma or opportunity. Which should it be?

There is generally a dichotomy between those living above and below a dam. Most if not all benefits and newly created opportunities seem to flow downstream, whereas the costs of submergence, displacement, and other disruption are left to be borne by marginalized catchment area communities. The directly affected population may be compensated and relocated; but what of all those others who remain? Since governments are committed to poverty alleviation and social justice, why not use the triggering mechanism of large dam projects to transform their lives as well?

Take Pancheswar. Its construction will give an impetus to the area development of the backward far-west region of Nepal and Pithoragarh district in the Kumaon region of Uttar Pradesh. With the opening up of these hitherto sequestered hinterlands and the establishment of viable market links, catchment area treatment could shade into designing new and more eco-friendly land use and productive cropping patterns. Highly angular slopes and unirrigated terraces could be put under tree crops rather than grain, and land use gradually diverted towards the development of horticulture, vegetable farming, herbiculture, floriculture, and alpine dairying with stall-fed animals.

The complementarity of growing seasons between hill and plain offers a nice market fit. With the opening up of communications, Himalayan villages are assured of food security should they produce "winter" vegetables, instead of cereals in the summer, when supplies are scarce in the plains and prices are high. Disease-free vegetable seeds grown at alpine heights would in turn find a ready market for winter sowing in the plains. Cheap energy could power ropeways to carry hill produce from more distant valleys to market hubs and agro-processing centres. Disciplined and trained ex-servicemen on either side of the border could constitute the nuclei of eco-development task forces to spearhead change.

This hydro-hardware programme could mesh with a parallel Mahakali basin human and area development programme in terms of education, health, nutrition, employment, and gender justice. A baseline quality-of-life survey could set minimal indices for social development which should be sought to be achieved within a given period. Such a programme would arrest and in some cases locally reverse the steady migration of people from degraded hills to the plains in search of employment.

Such a programme could be implemented by a Mahakali catchment area authority set up on either side of the border, initially financed by project funds for mandated activities such as catchment area treatment and rehabilitation. Later, a small upper-catchment development tax could be levied on the electricity generated to yield a steady income for the rest of the programme. Rehabilitation, in these circumstances, could be largely *in situ* rather than in distant locations, with the gestation period being used to educate and train the community to avail themselves of the new job opportunities planned to be created.

Wider regional cooperation

In the case of Indo-Bangladesh water resource development, too, while the direct benefits to both sides from specific projects and agreements are obvious, there is surely a larger purpose and more generous dividend to be gained. Water resource development is no more than an entry point to the bigger issue of alleviating poverty, hunger, and unemployment and leveraging a better quality of life for the large concentration of the world's poorest people who crowd the GBM basin. This too must be seen as a grand area development programme embracing a vibrant agriculture, the fashioning of new transport corridors and inter-modal carriage, port development, new patterns of hydro-electric/hydrocarbon/coal energy exchange, and the creation of new market opportunities and trans-border linkages. Optimization and synergy could be built into these loops.

Thus Chittagong could be a great entrepôt at the head of the Bay of Bengal, with Calcutta becoming a hub for inland waterways up the Ganges, Brahmaputra, and Barak. The restoration of connectivity between the north-east and the Indian heartland through Bangladesh and from all of this region to South-East Asia and south-west China through Myanmar could cement new regional relationships.

The World Bank and ADB have shown interest in subregional cooperation within the GBM basin, as has Thailand, which has promoted the BIMSTEC quadrant consisting of Bangladesh, India, Sri Lanka, and Thailand, with Mynamar lined up to join the grouping as well.

Within the GBM region, the Chakma settlement negotiated by Bangladesh in the Chittagong hill tract invites the formation of a larger Mizoram-Tripura-CHT economic subregion, with inland waterways along the Karnaphuli and Feni rivers linking Mizoram and Tripura to the world beyond through Chittagong port. Chitagong needs connectivity and a large hinterland to markets beyond.

North-east passage

The north-east, above all, requires transit and market opportunities. Its isolation can be greatly eased by regional cooperation, especially with Bangladesh.

The region has vast hydro-electric resources which would be attractive for private investment to exploit provided the state electricity boards/departments are corporatized, granted autonomy, and made subject to an independent regulatory authority that would set tariffs at economic levels. It has modest hydrocarbon and coal reserves as well, while Bangladesh appears poised to discover very large gas deposits. A regional energy grid and energy exchange would make sense.

Likewise, the Jamuna bridge across the Brahmaputra at Sirajganj in Bangladesh could become a nodal point for the projected Asian highway and railway, in which interest has revived. These would provide a land route for transit between the north-east and the Indian heartland on the one hand and South-East Asia and south-west China on the other. A national commission on basic minimum needs and infrastructural lags in the north-east (Planning Commission 1997) recommended a Rs 25,000 crores investment programme over the next decade and another Rs 75,000 crores during the ensuing 15 years. Bangladesh senses a rich market opportunity here.

Vision 2025

Water resource development is central to this vision, as it could bind a divided and fragmented region, become an engine for generating income and employment, and give a major thrust to infrastructure development and market opportunity. The Indian National Water Development Agency, under the guidance of a National Commission for Integrated

Water Resource Development, is looking at how best to manage the emerging water shortage, with inter-basin transfers where necessary by 2025. Bangladesh is engaged in a parallel exercise of optimizing its water resource development within the same time-scale through its Water Research and Planning Organisation, WARPO. Nepal's Electricity Development Corporation (EDC) is making a similar study with the focus on power. It would seem eminently desirable and logical that these three bodies establish contact, howsoever informally for a start, so as to commence a process of interaction. They have common goals and are dealing with a common resource.

Global warming and carbon trading

Environmental objections and dam safety issues remain. Certainly transparency, sustainability, and proper rehabilitation are pivotal. Some projects or aspects thereof may be clearly problematic. But not all large water resource projects are bad *per se*. Those that argue this case are ideologues who seek a different pathway to development and growth. For them, any and every large dam is conceptually undesirable. There is no countering doomsday views. Nothing is necessarily ideal. But the cost of not doing could be far greater environmentally and in human terms than damage from going ahead with due care and appropriate compensatory investments.

After swinging away from large dams, international opinion has begun to grope towards what could be a more stable and reasoned position. Global warming has made some earlier critics come round to the view that it is better for large developing societies like China and India to generate more clean hydro energy than to depend increasingly on fossil fuels, especially if used in dirty coal-fired thermal stations. The idea of carbon trading has been mooted as a means of reducing greenhouse gases. With proper safeguards, this could be one way of financing some of the Himalayan hydro projects under discussion.

From dialogue to doing

Both regionally and globally, therefore, the omens are propitious for proceeding energetically with developing the huge potential of the GBM basin. For those living in the region, the Ganges and Mahakali agreements hold out real hope for a brighter future. A framework has been fashioned. Now is the time to move from dialogue to action.

REFERENCES

Jha, H. B., ed. 1996. *Mahakali Treaty: Implications for Nepal's Development.* Nepal: Foundation for Economic and Social Change, Lalitpur. (See in particular contributions by P. S. Rana on "Mahakali Treaty: Benefits to Nepal" and P. C. Lohani on "Mahakali Treaty: A Vision for the 21st Century".)

Ministry of Water Resources. 1972. *Bangladesh Land and Water Resources Study.* Dhaka: Ministry of Water Resources.

Planning Commission. 1997. *Transforming the North-East: Tackling Backlogs in Basic Minimum Services and Infrastructural Needs.* Report of High-Level Commission to the Prime Minister chaired by S. P. Shukla. New Delhi: Government of India.

Verghese, B. G. 1990. *Waters of Hope.* Oxford/New Delhi: IBH Publishing.

Verghese, B. G. 1996. "Towards an Eastern Himalayan Rivers Accord", in Asit Biswas and Tsuyoshi Hashimoto, eds, *Asian International Waters: From Ganges-Brahmaputra to Mekong.* New Delhi: Oxford University Press.

Verghese, B. G. and Ramaswamy R. Iyer, eds. 1993. *Harnessing the Eastern Himalayan Rivers.* New Delhi: Konarak Publishers.

Verghese, B. G. and Ramaswamy R. Iyer, eds. 1994. *Converting Water Into Wealth.* New Delhi: Konarak Publishers.

Appendix

Treaty between the Government of the People's Republic of Bangladesh and the Government of the Republic of India on Sharing of the Ganga/Ganges Water at the Farakka

THE GOVERNMENT OF THE PEOPLE'S REPUBLIC OF BANGLADESH AND THE GOVERNMENT OF THE REPUBLIC OF INDIA,

DETERMINED to promote and strengthen their relations of friendship and good neighbourliness,

INSPIRED by the common desire of promoting the well-being of their peoples,

BEING desirous of sharing by mutual agreement the water of the international rivers flowing through the territories of the two countries and of making the optimum utilization of the water resources of their region in the fields of flood management, irrigation, river basin development and generation of hydro-power for the mutual benefit of the peoples of the two countries,

RECOGNIZING that the need for making an arrangement for sharing of the Ganga/Ganges waters at Farakka in a spirit of mutual accommodation and the need for a solution to the long-term problem of augmenting the flows of the Ganga/Ganges are in the mutual interests of the peoples of the two countries,

BEING desirous of finding a fair and just solution without affecting the rights and entitlements of either country other than those covered by this Treaty, or establishing any general principles of law or precedent,

HAVE AGREED AS FOLLOWS:

Article – I

The quantum of waters agreed to be released by India to Bangladesh will be at Farakka.

Article – II

(i) The sharing between India and Bangladesh of the Ganga/Ganges waters at Farakka by 10-day periods from the 1st January to the 31st May every year will be with reference to the formula at Annexure I and an indicative schedule giving the implications of the sharing arrangement under Annexure I is at Annexure II.
(ii) The indicative schedule at Annexure II, as referred to in sub-para (i) above, is based on 40 years (1949–1988) 10-day period average availability of water at Farakka. Every effort would be made by the upper riparian to protect flows of water at Farakka as in the 40 years average availability as mentioned above.
(iii) In the event flow at Farakka falls below 50,000 cusecs in any 10-day period, the two Governments will enter into immediate consultations to make adjustments on an emergency basis, in accordance with the principles of equity, fair play and no harm to either party.

Article – III

The waters released to Bangladesh at Farakka under Article I shall not be reduced below Farakka except for reasonable uses of water, not exceeding 200 cusecs, by India between Farakka and the point on the Ganga/Ganges where both its banks are in Bangladesh.

Article – IV

A Committee consisting of representatives nominated by the two Governments in equal numbers (hereinafter called the Joint Committee) shall be constituted following the signing of this Treaty. The Joint Committee shall set up suitable teams at Farakka and Hardinge Bridge to observe and record at Farakka the daily flows below Farakka Barrage, in the Feeder Canal, and at the Navigation Lock, as well as at the Hardinge Bridge.

Article – V

The Joint Committee shall decide its own procedure and method of functioning.

Article – VI

The Joint Committee shall submit to the two Governments all data collected by it and shall also submit a yearly report to both the Governments. Following submission of the reports the two Governments will meet at appropriate levels to decide upon such further actions as may be needed.

Article – VII

The Joint Committee shall be responsible for implementing the arrangements contained in this Treaty and examining any difficulty arising out of the implementation of the above arrangements and of the operation of Farakka Barrage. Any difference or dispute arising in this regard, if not resolved by the Joint Committee, shall be referred to the Indo-Bangladesh Joint Rivers Commission. If the difference or dispute still remains unresolved, it shall be referred to the two Governments, which shall meet urgently at the appropriate level to resolve it by mutual discussion.

Article – VIII

The two Governments recognize the need to cooperate with each other in finding a solution to the long-term problem of augmenting the flows of the Ganga/Ganges during the dry season.

Article – IX

Guided by the principles of equity, fairness and no harm to either party, both the Governments agree to conclude water-sharing Treaties/Agreements with regard to other common rivers.

Article – X

The sharing arrangement under this Treaty shall be reviewed by the two Governments at five years' interval or earlier, as required by either party and needed adjustments, based on principles of equity, fairness, and no harm to either party made thereto, if necessary. It would be open to either party to seek the first review after two years to assess the impact and working of the sharing arrangements as contained in this Treaty.

Article – XI

For the period of this Treaty, in the absence of mutual agreement of adjustments following reviews as mentioned in Article X, India shall release downstream of Farakka Barrange, water at a rate not less than 90% (ninety per cent) of Bangladesh's share according to the formula referred to in Article II, until such time as mutually agreed flows are decided upon.

Article – XII

This Treaty shall enter into force upon signature and shall remain in force for a period of thirty years and it shall be renewable on the basis of mutual consent.

IN WITNESS WHEREOF the undersigned, being duly authorized thereto by the respective Governments, have signed this Treaty.

DONE at New Delhi 12th December 1996 in Hindi, Bangla and English language. In the event of any conflict between the texts, the English text shall prevail.

Signed
(SHEIKH HASINA)
PRIME MINISTER,
PEOPLE'S REPUBLIC OF BANGLADESH

Signed
(H. D. DEVE GOWDA)
PRIME MINISTER,
REPUBLIC OF INDIA

Ganges Treaty: Annexure I

Availability at Farakka	*Share of India*	*Share of Bangladesh*
70,000 cusecs or less	50%	50%
70,000–75,000 cusecs	Balance of flow	35,000 cusecs
75,000 cusecs or more	40,000 cusecs	Balance of flow

Subject to the condition that India and Bangladesh each shall receive guaranteed 35,000 cusecs of water in alternate three 10-day periods during the period 1 March to 10 May.

Ganges Treaty: Annexure II

Sharing of waters a Farakka between 1 January and 31 May every year

If actual availability corresponds to average flows of the period 1949 to 1988, the implication of the formula in Annexure I for the share of each side is as shown below.

Period	Average of total flow 1949–1988 (cusecs)	India's share (cusecs)	Bangladesh's share (cusecs)
January			
1–10	107,516	40,000	67,516
11–20	97,673	40,000	57,673
21–31	90,154	40,000	50,154
February			
1–10	86,323	40,000	46,323
11–20	82,859	40,000	42,859
21–28	79,106	40,000	39,106
March			
1–10	74,419	39,419	35,000
11–20	68,931	33,931	35,000*
21–31	64,688	35,000*	29,688
April			
1–10	63,180	28,180	35,000*
11–20	62,633	35,000*	27,633
21–30	60,992	25,992	35,000*
May			
1–10	67,351	35,000*	32,351
11–20	73,590	38,590	35,000
21–31	81,854	40,000	41,854

* Three 10-day periods during which 35,000 cusecs shall be provided

Acronyms

ADB	Asian Development Bank
APP	Agriculture Perspective Plan (Nepal)
bcm	billion cubic metres
BIWTA	Bangladesh Inland Waterways Transport Authority
BWDB	Bangladesh Water Development Board
CEA	Central Electricity Authority (India)
CNRET	Centre for Natural Resources, Energy, and Transport (UN Department of Economic and Social Affairs)
DIWT	Department of Inland Water Transport (India)
DPR	detailed project report
EHV	extra-high voltage
FMIS	farmer-managed irrigation systems
GBM	Ganges-Brahmaputra-Meghna
GDA	Ganges-dependent area
GDP	gross domestic product
GNP	gross national product
HV	high voltage
IIMI	International Irrigation Management Institute
ILA	International Law Association
ILC	International Law Commission
JRC	Joint Rivers Commission (Indo-Bangladesh)
LUPP	land-use planning project (Bhutan)
NEA	Nepal Electricity Authority

PSMP	power system master plan (Bhutan)
SAARC	South Asian Association for Regional Cooperation
SPARRSO	Space Research and Remote Sensing Organization
TDR	transferable development rights
UNDP	United Nations Development Programme
UNEP	United Nations Environment Programme
UNESCO	United Nations Educational, Scientific, and Cultural Organization
WARPO	Water Research and Planning Organization (Bangladesh)
WHO	World Health Organization
ZACPLAN	Zambesi action plan

List of contributors

Asit K. Biswas is the president of the Third World Centre for Water Management and a member of the World Commission on Water. A former president of the International Water Resources Association, his work has been translated into 31 languages.

A. T. M. Shamsul Huda is the secretary of the Ministry of Water Resources of the government of Bangladesh, and has guided and shaped the country's water policies for nearly a decade.

Ratneshwar Lal Kayastha is the joint secretary of the Ministry of Water Resources of the government of Nepal.

A. D. Mohile is the former chairman of the Brahmaputra Board, and the current chairman of the Central Water Commission of the government of India.

Somnath Mukherjee is the head of the Environment Division of WAPCOS, New Delhi, a public-sector consulting group.

Ainun Nishat is the country representative of IUCN in Bangladesh, and was earlier a professor of water resources engineering at the Bangladesh University of Science and Technology in Dhaka.

Iswer R. Onta was instrumental in establishing Eastconsult, which is Nepal's best-known consulting group in the area of water and irrigation.

R. B. Shah is the former chairman of the Central Water Commission of the government of India, and currently an adviser to Consulting Engineering Services of New Delhi.

Juha I. Uitto is a monitoring and evaluation specialist at the Global Environment Facility (GEF), and was formerly senior academic programme officer at the United Nations University Environment and Sustainable Development Programme.

B. G. Verghese is a professor at the Centre for Policy Research, New Delhi, and a former editor of the *Indian Express* and the *Hindusthan Times*.

Index